# MEDIEVAL SCIENCE AND TECHNOLOGY

BIBLIOGRAPHIES OF THE HISTORY
OF SCIENCE AND TECHNOLOGY
(Vol. 11)

GARLAND REFERENCE LIBRARY
OF THE HUMANITIES
(Vol. 494)

# Bibliographies of the History of Science and Technology

*Editors*



1. *The History of Modern Astronomy and Astrophysics*
   *A Selected, Annotated Bibliography*
   by David DeVorkin
2. *The History of Science and Technology in the United States*
   *A Critical, Selective Bibliography*
   by Marc Rothenberg
3. *The History of the Earth Sciences*
   *An Annotated Bibliography*
   by Roy Sydney Porter
4. *The History of Modern Physics*
   *An International Bibliography*
   by Stephen Brush and Lanfranco Belloni
5. *The History of Chemical Technology*
   *An Annotated Bibliography*
   by Robert P. Multhauf
6. *The History of Mathematics From Antiquity to the Present*
   *A Selective Bibliography*
   by Joseph Dauben, editor
7. *The History of Meteorology and Geophysics*
   *An Annotated Bibliography*
   by Stephen G. Brush and Helmut E. Landsberg
8. *The History of Classical Physics*
   *A Selected, Annotated Bibliography*
   by R. W. Home, with assistance of Mark J. Gittins
9. *The History of Geography*
   *An Annotated Bibliography*
   by Gary S. Dunbar
10. *The History of the Health Care Sciences and Health Care, 1700–1980*
    *A Selective Annotated Bibliography*
    by Jonathon Erlen
11. *Medieval Science and Technology*
    *A Selected, Annotated Bibliography*
    by Claudia Kren

# MEDIEVAL SCIENCE AND TECHNOLOGY
## *A Selected, Annotated Bibliography*

Claudia Kren

GARLAND PUBLISHING, INC. • NEW YORK & LONDON
1985

**Library of Congress Cataloging in Publication Data**

Kren, Claudia, 1929–
Medieval science and technology.

(Bibliographies of the history of science and technology ; v. 11) (Garland reference library of the humanities ; v. 494)
Includes index.
1. Science, Medieval—Bibliography. 2. Technology—History—Bibliography. I. Title. II. Series. III. Series: Garland reference library of the humanities ; v. 494.
Z7405.H6K74 1985 [Q124.97] 016.5′09′02 84-48012
ISBN 0-8240-8969-3 (alk. paper)

Printed on acid-free, 250-year-life paper
Manufactured in the United States of America

For Anne G. Edwards

# GENERAL INTRODUCTION

This bibliography is one of a series designed to guide the reader into the history of science and technology. Anyone interested in any of the components of this vast subject area is part of our intended audience, not only the student, but also the scientist interested in the history of his own field (or faced with the necessity of writing an "historical introduction") and the historian, amateur or professional. The latter will not find the bibliographies "exhaustive," although in some fields he may find them the only existing bibliographies. He will in any case not find one of those endless lists in which the important is lumped with the trivial, but rather a "critical" bibliography, largely annotated, and indexed to lead the reader quickly to the most important (or only existing) literature.

Inasmuch as everyone treasures bibliographies, it is surprising how few there are in this field. George Sarton's *Guide to the History of Science* (Waltham, Mass., 1952; 316 pp.), Eugene S. Ferguson's *Bibliography of the History of Technology* (Cambridge, Mass., 1968; 347 pp.), François Russo's *Histoire des Sciences et des Techniques. Bibliographie* (Paris, 2nd ed., 1969; 214 pp.) are justifiably treasured but they are, of necessity, limited in their coverage and need to be updated.

For various reasons, mostly bad, the average scholar prefers adding to the literature to sorting it out. The editors are indebted to the scholars represented in this series for their willingness to expend the time and effort required to pursue the latter objective. Our aim, and that of the publisher, has been to give the series enough uniformity to give some consistency to the series, but otherwise to leave the format and contents to the author/compiler. We have urged that introductions be used for essays on "the state of the field," and that selectivity be exercised to limit the length of each volume. Since the historical literature ranged from very large (e.g., medicine) to very small (e.g., chemical technology), some bibliographies will be limited to truly important writings while others will include modest "contributions" and even primary sources. The problem is intelligible guidance into a particular field—or subfield—and its solution is largely left to the author/compiler. In general, topical volumes (e.g., chemistry) will deal with the subject since about 1700, leaving earlier literature to area or chronological volumes (e.g., medieval science); but here, too, the volumes will vary according to the judgment of the author. The volumes are international (except for two, *Science and Technology in the United States* and *Science and Technology in Eastern Asia*) but the literature covered depends, of course, on the linguistic equipment of the author and his access to "exotic" literatures.

Robert Multhauf
Ellen Wells

*Smithsonian Institution*
*Washington, D.C.*

# CONTENTS

# INTRODUCTION

Early in this century, if not before, beginning with the initial researches of a few pioneers, the history of medieval science attracted a steadily increasing number of scholars who, with their editions of texts and their interpretations of medieval scientific thought, successfully established the field as an important research area. Moreover, as a result of these endeavors, medieval science has now gained recognition as an autonomous enterprise, worthy of study in its own right within the context of the culture which nurtured it, and not simply investigated for possible adumbrations of the scientific developments of the early modern period.

Over the past three decades in particular, the amount of scholarship devoted to all areas of medieval science and technology has increased considerably. The number of critical editions of medieval texts, often accompanied by translation into a modern language, and articles in learned journals of all kinds, not just those concerned with the history of science, serve as only two indexes of the vitality of the field. Given the amount of the literature now available in print, it is the purpose of this bibliography to gather within one volume a selection of this scholarship from the earliest period through recent years, although the bibliography is indeed select and is not intended as an exhaustive listing of the work of any scholar whether medieval or modern.

Several types of material have been deliberately excluded. Reviews, dissertations, or surveys of any area in the history of science lacking substantial medieval treatment will not be found here. Also, despite a long tradition as a quadrivial art and its place within the arts curriculum of many medieval universities, music, as a distinct category, has been omitted, although materials which treat music in connection with, say, medicine or natural philosophy, have been included in the appropriate place. More significantly, perhaps, with the exception of a few editions of such works by Ockham (items 342, 347, 349), theological treatises, especially commentaries on the *Sentences* of Peter Lombard and collections of quodlibetal questions, are not listed here even though some are available. The omission of these materials is certainly not indicative of their importance for the study of medieval scientific thought—they include opinion on a wide spectrum of topics, from the nature of motion to astronomical theories. Several of the entries in this bibliography reveal not only that theological works can be mined for their scientific content but also that the contexts in which these notions are embedded evidence the link between theological concerns and medieval natural philosophy. This point is illustrated and elaborated, especially in items 336, 360, 363, 364, 669a, 671, 1077. A guide to manuscripts and early editions of commentaries on the *Sentences* can be found in Stegmüller (item 49), while Glorieux (item 27) has similar information on the quodlibetal literature. Several early editions of theological treatises have been reprinted by Minerva and Gregg, for example, while the Franciscan Institute, St. Bonaventure, N.Y., is now publishing modern editions of Ockham's theological works. (The Institute has also reprinted the 1522 edition of the *Sentences* commentary of Gregory of Rimini.)

The bibliography covers primarily western Europe, or more exactly, the cultural entity known as the medieval Latin West, from approximately the fifth century to circa 1450. There are occasional exceptions to this time frame. It includes materials on and by Arabic (and Jewish) scholars whose works entered Christian Europe in Latin translation and became an integral part of the medieval scientific and philosophical tradition. Literature in the major western European languages has been surveyed through 1981 (with some materials from 1982); there is no lower time limit here. Although the time period included in the bibliography is extremely long, the organization is, on the whole, not chronological but thematic, the sections and subsections being the various sciences themselves and related areas.

There are three types of entry in the bibliography: first, editions of medieval texts, both modern critical editions as well as reprints of early editions made available by the reprint publishers; second, modern editions of medieval translations of the authoritative sources of medieval scientific and philosophical thought, the scholarship resulting from several on-going projects, for example, *Aristoteles latinus*. Materials of these first two types have either not been annotated or annotated very briefly; it is assumed in such cases that the content is known and it is the bibliographical information which is of use. The third kind of entry comprises the secondary literature—books, articles and essays in collections. These materials have been annotated extensively. Throughout the divisions of the bibliography, no distinction has been made among these varieties of materials, and any section may contain examples of all types.

There are a number of cross-references in the bibliography, especially where matters of ascription or interpretation are involved. A few items which very obviously might be placed under more than one heading have been entered twice, the second listing indicated by an asterisk. But this double-listing device has been sparingly employed. Given the unitary character of medieval scientific culture, the multi-listing which would be useful in a bibliography on the scientific thought of a later period would be an invitation to chaos here. Within the bibliography, the names of medieval authors have been alphabetized by either first or last name depending on which is more usually employed. However, perceptions concerning this vary, and it would be well to look up any medieval writer under both names. Arabic names have been given in the form adopted in the Latin West, where this is applicable, followed by the Arabic version in parentheses.

As mentioned before, the classification scheme of the bibliography is for the most part thematic, although chronology is tacit in several of its sections and in one instance has been explicitly introduced. The kind of materials one would expect to find subsumed under a particular rubric in many of the divisions is readily apparent—for example, there is no ambiguity as to what might be located under *Astronomical Instruments, Sundials, and Clockwork*. However, certain of the sections do require some explanation as to what they contain or, for that matter, what they do not.

## General Research Aids

Among the various bibliographies, reference works and collections of documents included here, there has been no attempt to list manuscript catalogues, although a few specialized lists or catalogues of manuscript materials have been provided. Materials of that kind are given in Kristeller (item 33), and the indexes to the catalogues in Kristeller are now available on microfilm (item 38).

Research in any aspect of medieval science or culture is dependent on the availability of manuscript materials. While repositories abroad or in the United States, certainly the larger ones, provide microfilm or paper reproductions of materials on request, it seems appropriate to mention here a few large microfilm collections in this country which not only provide this service but also welcome scholars. For example, the Vatican Film Library at St. Louis University, St. Louis, Missouri, which has microfilm copies of the Vatican collections, the Hill Monastic Manuscript Library, St. John's University, Collegeville, Minnesota, which began by filming the contents of the Austrian monasteries and the National Library at Vienna, but which now has filmed many other collections, and the Mediaeval Institute at the University of Notre Dame, which has the Ambrosiana at Milan on microfilm (item 26).

Likewise while not listed in the bibliography, the great national collections of printed documents are a valuable source for the historian of medieval science. Many of these, such as the *Monumenta Germaniae historica* or the *Rerum Britannicarum medii aevi scriptores*, with full bibliographical information, can be found in Paetow (item 28).

There are also projects or sources, some now in preparation, that can or will be able to provide information concerning the content or location of scientific manuscripts. For example, the services offered by the Centre National de la Recherche Scientifique (including its bibliographical volumes, the CNRS *Bulletin signalétique*, 522: *Histoire des sciences et des techniques*. Projects now under way include the Benjamin Data Bank of Medieval Scientific Manuscripts in Latin which will be able to provide scholars with a retrieval service drawn from an extensive data bank of descriptions of manuscripts. Inquires concerning this project should be addressed to Nan L. Hahn, Rutgers University, New Brunswick, N.J. Menzo Folkerts of the Institut für Geschichte der Naturwissenschaften at the University of Munich is likewise directing a mathematical manuscript project (items 1417, 1418).

## Encyclopedic Tradition

This section contains not only works by and materials on the early medieval authors traditionally included in the encyclopedic and handbook tradition (Cassiodorus, Isidore and Macrobius, for example), but also authors of large-scale works in the Carolingian Age (Erigena) and of the twelfth century (Adelard of Bath, Honorius of Autun, for example). The encyclopedia as a genre did not vanish with the advent of Greek and Arabic scientific materials; late-twelfth-century and thirteenth century (and beyond) encyclopedists such as Thomas of Cantimpré and Vincent of Beauvais are also included.

## Physics (Optics Excluded) and Natural Philosophy Before the Fourteenth Century

Broadly speaking, this subdivision is designed to reflect aspects of physical thought, excluding optics and mathematical as opposed to physical astronomy, of the thirteenth century. It includes translations of texts and commentaries on Greek and Arabic authors on the *libri naturales* of Aristotle (*Physics, De caelo et mundo, De generatione et corruptione* and *Meteorology*) which became available in Latin in this century as well as the commentaries on these works by thirteenth-century schoolmen. It likewise includes material on a variety of topics in physical science, for example, kinematics, statics, magnetism, tides, again, overwhelmingly of the thirteenth century.

## Natural Philosophy of the Fourteenth Century

The natural philosophy of this century (and beyond) has been explicitly distinguished from that of the previous period so as to emphasize the creativity of fourteenth-century scholastics who reinterpreted or transcended ancient authority and who developed new conceptual tools and methodologies in the treatment of physical problems, what John Murdoch has called "new analytic languages." (On this theme, see especially items 339, 669, 669a, 670, 671.)

## Astronomy and Cosmology

The descriptive physical astronomy and cosmology the Middle Ages inherited from the Eudoxian-Aristotelian tradition, usually explicated in commentaries on the *De caelo* and the *Metaphysics*, has been listed elsewhere. This sub-section contains materials on cosmology not directly derived from Aristotelian commentaries, on astronomical observations; above all, it lists texts and materials which deal with the technical, mathematical astronomical tradition that scholastics owed directly or indirectly to Ptolemy and Arabic authors.

## Philosophy and Metaphysics

This subdivision contains a selection of medieval philosophical works relevant to the study of nature; it provides a location for translations of, and commentaries on, the *Metaphysics* and includes materials on doctrinal disputes, for example, on Averroism, theories of knowledge and the eternity of the world.

## Geometry, Practical Geometry, Proportions, Conic Sections, Trigonometry

Works on proportions given here treat this topic within the context of its Greek and Arabic origins. The medieval doctrine of proportions as an "analytic language," that is, applied to physical problems, is included under Natural Philosophy of the Fourteenth Century.

## Medicine

No effort has been made in this section to list the considerable literature dealing with various aspects of the Black Death and its impact on medieval society. Entries on this subject are confined to Sudhoff's extensive collection of plague tracts (item 976) and Ziegler's *The Black Death,* which has an excellent bibliography (item 992).

## Psychology

This section includes, among other materials, texts of and commentaries on the *De anima* and the *Parva naturalia* as well as discussions of problems arising from these works. It also includes entries on the topic of sensible species in the medium although this was not entirely a psychological problem.

## Education: Schools and Universities

Included here are materials on the origins and development of medieval schools and universities, the organization of faculties, on curricula, on texts employed and teaching methods. But with the exception of two collections of documents stemming from the University of Paris (items 1224, 1257), no effort has been made to list the large number of university documents now in print, for example those on Vienna published by the Institut für Österreichische Geschichtsforschung or the many documents concerning British universities reprinted by Kraus. A guide to such collections as well as to standard histories of medieval universities (although not the most recent ones) is provided in the bibliographical introductions prefacing treatment of individual universities in Rashdall (item 1209).

## Quasiscience

Here are listed materials on astrology, alchemy and various forms of magic and divination, medieval enterprises most often classified as "pseudo-science." However, a sympathetic consideration of the stature of these subjects within the context of the physical, cosmological and theological suppositions of medieval culture would surely find this appellation too anachronistic. But certainly, astrology and alchemy, much less magical practices, hardly enjoyed a universal approbation among scholastics; hence the rubric "quasi (as if) science."

## Reports on Manuscripts

Many modern scholars engaged with medieval science have often provided lists of manuscripts of works by specific authors or have reported on collections of such materials dealing with particular themes, usually with no more than brief summaries of their contents, or have announced the discovery of new manuscript versions or adduced new evidence concerning attribution. A selection of this kind of material is included here; admittedly, the distinction between some of these reports and a research aid is exiguous.

I am glad to have an opportunity to acknowledge the assistance I have received from many people in the compilation of this bibliography, especially from the staffs of the libraries of Health Sciences, Veterinary Science, of the Mathematics and Engineering Libraries as well as of the inter-library loan services of the Ellis Library, all of the University of Missouri-Columbia. I am also indebted to Dr. Thomas W. Shaughnessy, Director of the Ellis Library, for his kindly aid, to Gerry Burke, who with patient skill typed a difficult manuscript, to the Research Council of the University of Missouri-Columbia for financial assistance and to the editors-in-chief of this series of bibliographies for their valuable suggestions. Above all, I owe much to the cheerful perseverance and ready expertise of the Ellis Library's history librarian, Anne E. Edwards.

# Medieval Science and Technology

# GENERAL

## GENERAL WORKS AND *OPERA OMNIA*


1. Albertus Magnus. *Opera omnia*, 38 vols. Edited by August Borgnet. Paris: Vivès, 1890-99.

2. Alfarabi (al-Fārābī). *Opera omnia quae latina lingua conscripta reperiri potuerunt*. Edited by G. Camerarius. Paris, 1638. Reprint. Frankfort: Minerva, 1969.

3. Averroes (Ibn Rushd). *Aristotelis omnia quae extant opera cum Averrois commentariis*. 9 vols. in 14. Venice, 1562-1574. Reprint. Frankfort: Minerva, 1962.

4. Avicenna (Ibn Sina). *Avicenne perhypatetici philosophi....* 1508. Reprint. Frankfort: Minerva, 1961.

5. Bacon, Roger. *Opera hactenus inedita*. 16 vols. in 14. Edited by Robert Steele. Oxford: Clarendon Press, 1909-1940.

6. Bacon, Roger. *Opera quaedam hactenus inedita*. 3 vols. Edited by J.S. Brewer. London: Longman, Green, Longman & Roberts, 1959.

   Contains the *Opus tertium*, *Opus minus* and the *Compendium philosophiae*.

7. Beaujouan, Guy. "Medieval Science in the Christian West." *History of Science*, I: *Ancient and Medieval Science from the Beginnings to 1450*. Edited by René

Taton. Translated by A.J. Pomerans. New York: Basic Books, 1963, pp. 468-531.

Surveys the major topics in medieval science divided chronologically into four periods: (1) Dark Ages, fifth through tenth centuries; (2) the awakening of Europe and the reception of Arabic sciences, eleventh through twelfth centuries; (3) the rise of the university and the high point of scholastic science; (4) the decline of the university and the transition of science into technology. Translated from René Taton. Histoire général des sciences, vol. I. Paris: Presses universitaires de France, 1957.

8. Cockayne, Thomas O. Leechdoms, Wortcunning, and Starcraft of Early England. 3 vols. London, 1864-1866. Reprint. Rev. ed. with an introduction by Charles Singer. London: Holland Press, 1961.

A collection of documents dealing with approaches to nature in Anglo-Saxon England.

9. Crombie, A.C. Medieval and Early Modern Science. 2 vols. Rev. 2d ed. New York: Doubleday Anchor, 1959.

A revision of Augustine to Galileo: The History of Science A. D. 400-1650, London, 1952. The first volume covers science from the Vth-XIIIth centuries; volume two, science in the later M. A. and early modern times, XIIIth-XVIIth centuries. A survey of medieval scientific thought which emphasizes the continuity of the scientific tradition in the West from the Greeks to the seventeenth century.

10. Grant, Edward, ed. A Source Book in Medieval Science. Cambridge, Mass.: Harvard University Press, 1974. Pp. xviii + 864.

Includes representative source readings drawn from medieval authors from the Latin encyclopedists to the fifteenth century, and on occasion, other materials. There are approximately 190 selections covering a broad spectrum of medieval scientific work including astrology and alchemy; technology was deliberately ommitted. Informative introductions, extensive annotations, useful cross-references and scholarly notes accompany most of the selections. The volume concludes with a section on author biographies.

11. Gunther, R.T. Early Science in Oxford. 15 vols. Oxford: Oxford University Press, 1920-1940.

Of special interest are: Vol I-chemistry, mathematics, surveying. Vol II-astronomy. Vol. III-biology. Vol. V-Chaucer and Messahalla on the astrolabe.

12. Gunther, R.T. Early Science in Cambridge. Oxford, 1937. Reprint. London: Dawsons of Pall Mall, 1964. Pp. xii + 513.

While not primarily concerned with the medieval period, contains some material on medieval mensuration, medicine and astronomical instruments.

13. Lindberg, David C., ed. Science in the Middle Ages. Chicago: University of Chicago Press, 1978. Pp. xv + 549.

A collection of essays on major aspects of medieval science, culture and learning by Brian Stock, "Science, Technology, and Economic Progress in the Early Middle Ages," pp. 1-51: David C. Lindberg, "The Transmission of Greek and Arabic Learning to the West," pp. 52-90 and "The Science of Optics," pp. 338-368; William A. Wallace, "The Philosophical Setting of Medieval Science," pp. 91-119; Pearl Kibre and Nancy G. Siraisi, "The Institutional Setting; The Universities," pp. 120-144; Michael S. Mahoney, "Mathematics," pp. 145-178; Joseph E. Brown, "The Science of Weights," pp. 179-205; John E. Murdoch and Edith D. Sylla, "The Science of Motion," pp. 206-264; Edward Grant, "Cosmology," pp. 265-302; Olaf Pedersen, "Astronomy," pp. 303-337; Robert P. Multhauf,"The Science of Matter," pp. 369-390; Charles H. Talbot, "Medicine," pp. 391-428; Jerry Stannard, "Natural History," pp. 429-460; James A. Weisheipl, "The Nature, Scope and Classification of the Sciences," pp. 461-482; Bert Hansen, "Science and Magic," pp. 483-506.

14. Nasr, Seyyed Hossein. Science and Civilization in Islam. New York: New American Library, 1970. Pp. xiv + 384.

Surveys Islamic contributions to the classification of the sciences, astronomy and cosmology, geography, natural history, physics, mathematics, medicine, philosophy and alchemy. The thought of many of the individuals discussed became part of the scientific and philosophical tradition of the Latin West.

15. Scotus, John Duns. Opera omnia. 17 vols. Studio et cura commissionis Scotisticae. Vatican City: Typis polyglottis Vaticanis, 1950-1966.

16. Scotus, John Duns. Opera omnia. 12 vols. in 16. Lyon, 1639. Hildesheim: Olms, 1968-69.

A reprint of the Wadding edition.

17. Stahl, William H. Roman Science: Origins, Development and Influence to the Later Middle Ages. Madison: University of Wisconsin Press, 1962. Pp. vii + 308.

Traces the origins of Roman thought in classical Greece and the early Hellenistic period, its development in the western Empire to the fifth century and the work of Macrobius and Martianus Capella. Surveys encyclopedic science in Ostrogothic Italy (Boethius and Cassiodorus), and its use in the writings of Isidore and Bede, in the Carolingian world and at Chartres.

18. Steneck, Nicholas H. Science and Creation in the Middle Ages: Henry of Langenstein (d. 1397) on Genesis. Notre Dame: University of Notre Dame Press, 1976. Pp. xiv + 213.

Surveys medieval science as understood by a late medieval schoolman revealed in his lectures on the beginning chapters of Genesis.

19. Thorndike, Lynn. A History of Magic and Experimental Science. 8 vols. New York: Columbia University Press, 1923-1958.

A monumental classic based on extensive use of MS materials. The first four volumes (Pliny through the fifteenth century) are relevant to the medieval period. The underlying thesis is that magic (the occult arts) and science (the experimental exploration of nature) rose together and retained their mutual interconnections in the Middle Ages.

20. William of Auvergne. Opera omnia. 2 vols. Paris, 1674. Reprint. Frankfort: Minerva, 1963.

RESEARCH AIDS

21. Beccaria, Augusto. I codici de medicina del periodo presalernitano (secoli IX, X, e XI). Rome: Edizioni di storia e letteratura, 1956. Pp. 500.

A catalogue of early medical MS materials in European collections arranged by country.

22. Bibliography of the History of Medicine. History of Medicine Division. Bethesda, MD: National Library of Medicine, 1965--.

An annual bibliography (cumulated every five years) of current materials on the history of medicine and related areas covering all periods. Vol. 18 appeared in 1982.

23. Carmody, Francis J. Arabic Astronomical and Astrological Sciences in Latin Translation. Berkeley: University of California Press, 1956. Pp. 193.

A chronological arrangement of works which were transmitted to the West through Arabic intermediaries, or Arabic treatises which became known to the Latins. Excludes commentaries on the De caelo and the Metaphysics as well as Latin materials based on Hebrew translations of Arabic works.

24. Cranz, F. Edward, ed.-in-chief. Catalogus translationum et commentariorum: Medieval and Renaissance Latin Translations and Commentaries. Washington, D.C.: Catholic University of America, 1960--.

A continuing series describing the Latin translations of Greek authors, including philosophy, science and pseudo-science, and the Latin commentaries on ancient Latin and Greek works up to 1600. Vol. IV appeared in 1980.

25. Dictionary of Scientific Biography. Edited by Charles C. Gillispie. 16 vols. New York: Charles Scribner's Sons, 1970-1980.

Contains entries on the lives, ideas and works of the major personalities in medieval scientific thought and natural philosophy.

26. Gabriel, Astrik L. A Summary Catalogue of Microfilms of One Thousand Scientific Manuscripts in the Ambrosiana Library, Milan. Notre Dame: Mediaeval Institute of Notre Dame, 1968. Pp. 439.

The Mediaeval Institute at Notre Dame has the Ambrosiana Library of Milan on microfilm. This catalogue of selected MSS covers the entire range of science and technology.

27. Glorieux, Palémon. La littérature quodlibétique. 2 vols. Paris: Le Saulchoir Kain, 1925; J. Vrin, 1935. Pp. 382, 387.

Presents in Vol. I a register of quodlibetal disputations at Paris for the period 1260-1320 with brief biographical notices, incipits and explicits of the Quodlibets and their question titles. Vol. II contains a further list, including Quodlibets whatever their provenance or date, and their MSS.

28. A Guide to the Study of Medieval History. Edited by Louis J. Paetow. 2d ed. New York, 1931. Reprint. New York: Kraus, 1980. Pp. cxii + 643.

Contains bibliographical information on collections of original sources, medieval scientific thought and universities.

29. Hauréau, Barthélemy. Notices et extraits de quelques manuscrits latins de la Bibliothèque Nationale. Paris, 1890-93. Reprint. 6 vols. in 3. Farnborough: Gregg, 1967.

30. International Medieval Bibliography. Edited by Richard J. Walsh. Leeds: W.S. Mauley & Son, 1968--.

A bibliography now published twice a year.

31. Isis Cumulative Bibliography. A Bibliography of the History of Science Formed from Isis Critical Bibliographies 1-90, 1913-1965. 5 vols. + Supplement. Edited by Magda Whitrow. London: Mansell, 1971--.

Vols. I and II: Personalities and Institutions, Vol. III: Subjects, Vol. IV: Civilizations and Periods-Pre-History to the Middle Ages, Vol. V: Civilizations and Periods-15th-19th Centuries. A supplementary volume

(Personalities and Institutions), published in 1980, was edited by John Neu.

32. Jacquart, Danielle. Le milieu médical en France du XIIe au XVe siècle. En annexe 2e supplément au Dictionnaire d'Ernest Wickersheimer. Geneva: Droz, 1981. Pp. 487.

See item 55.

32a. Klebs, Arnold C. "Incunabula scientifica et medica: Short Title List." Osiris, 4 (1938), 1-359.

An alphabetical list of about 650 authors.

33. Kristeller, Paul O. Latin Manuscript Books before 1600: A List of the Printed Catalogues and Unpublished Inventories of Extant Collections. 3d ed. New York: Fordham University Press, 1965. Pp. xxvi + 284.

A reference work in three parts: A bibliography of works on libraries and their collections of MSS, a list of works which describe MSS of more than one city and a list of printed catalogues and handwritten inventories of individual collections by cities.

34. Lacombe, George, A. Birkenmajer, M. Dulong, E. Franceschini and L. Minio-Paluello, eds. Aristoteles latinus. 2 vols. Rome, 1939, Cambridge, 1955. Editio nova. Bruges: Desclée de Brouwer, 1957--.

Provides information on translations of Aristotelian and pseudo-Aristotelian works and descriptions of MSS.

35. Lindberg, David C. A Catalogue of Medieval and Renaissance Optical Manuscripts. Toronto: Pontifical Institute of Mediaeval Studies, 1975. Pp. 142.

36. Literature of Medieval History, 1930-1975. Compiled and edited by Gray Cowan Boyce. 5 vols. New York: Kraus, 1981.

A bibliography with sections on medieval science and related areas. It is a supplement to Paetow's Guide to the Study of Medieval History. See item 28.

37. Markowski, Mieczysław and Sophia Włodek. Repertorium commentariorum medii aevi in Aristotelem latinorum quae in Bibliotheca Iegellonica Cracoviae asservantur. Warsaw: Ossolineum, 1974. Pp. 210.

Medieval Aristotelian commentaries in the Jagellonian Library at the University of Cracow.

38. A Microfilm Corpus of the Indexes to Printed Catalogues of Latin Manuscripts before 1600 A.D. Based on Paul O. Kristeller, Latin Manuscript Books Before 1600 A.D., 3d ed. Prepared under the direction of F. Edward Cranz with Paul O. Kristeller, 1982. Microfilm: 38 reels. Text: Pp. ix + 609.

Contains the indexes to authors and works from MS catalogues listed in Kristeller, Latin Manuscript Books before 1600, A.D., 3d ed. There is also an accompanying volume containing bibliographical information on the items on microfilm. Copies can be obtained from the Connecticut College Bookstore, Connecticut College, New London, Conn. See item 33.

39. Migne, J.P. Patrologiae cursus completus ... Series latina. 221 vols. in 222. Paris: Garnier, 1878-1890.

Contains documents by personalities who wrote on science and philosophy.

40. Millás Vallicrosa, J.M. Las traducciones orientales en los manucritos de la Biblioteca Catedral de Toledo. Madrid: Consejo superior de investigaciones científicos, 1942. Pp. 371.

A MS catalogue of Latin translations from Arabic and Hebrew in the cathedral library at Toledo, with texts of astronomical treatises in the appendixes: works on the astrolabe by Maslama, Massallah and John of Spain, and the astronomical treatise of Alhazen.

41. Minio-Paluello, Lorenzo, ed. Aristoteles latinus. Bruges: Desclée de Brouwer, 1961. Pp. 229.

Commentaries and bibliography.

42. Mohan, G.E. "Incipits of Logical Writings of the XIII-XVth Centuries." *Franciscan Studies*, 12 (1952), 349-389.

43. *New Catholic Encyclopedia*. 17 vols. Editorial Staff of Catholic University. New York: McGraw-Hill, 1967-1979.

Contains entries on personalities and topics related to medieval science, philosophy and theology.

44. Sarton, George. *Introduction to the History of Science*. 3 vols. Baltimore, MD: Williams & Wilkins, 1927, 1931, 1948.

A bio-bibliographical reference work covering the East and Europe from Homer to 1400.

45. Sarton, George. *Horus: A Guide to the History of Science*. Waltham, Mass.: Cronica Botanica, 1952. Pp. xvii + 316.

A still useful source, particularly for reference works.

46. Singer, Dorothea W. *Catalogue of Latin and Vernacular Alchemical MSS in Great Britain and Ireland Dating Before the XVI Century*. 3 vols. Brussels: M. Lamertin, 1928-31.

47. Smet, A.J. *Initia commentariorum questionum et tractatuum latinorum in Aristotelis libros de anima saeculis XIII, XIV, XV editorum*. Louvain: De Wulf, 1963. Pp. 105.

48. Spade, Paul V. *The Mediaeval Liar: A Catalogue of the Insolubilia-Literature*. Toronto: Pontifical Institute of Mediaeval Studies, 1975. Pp. 137.

49. Stegmüller, Friedrich. *Repertorium commentariorum in Sententias Petri Lombardi*. 2 vols. Würzburg: Schöningh, 1947.

Lists authors of *Sentences* commentaries with bibliographical information including MSS and editions.

50. Stillwell, Margaret Bingham. *The Awakening Interest in Science during the First Century of Printing, 1450-*

1550. New York: Bibliographical Society of America, 1970. Pp. xxix + 399.

An annotated list of first editions of scientific works in astronomy, mathematics, medicine, natural history, physics and technology.

51. Subject Catalogue of the History of Medicine and Related Sciences. 18 vols. Wellcome Institute for the History of Medicine, London. Munich: Kraus, 1980.

A comprehensive index to the secondary literature.

51a. A Summary Checklist of Medical Manuscripts on Microfilm Held by the National Library of Medicine. Compiled by Richard J. Durling. Bethesda, MD: National Library of Medicine, 1968. Pp. 14.

52. Suter, Heinrich. Die Mathematiker und Astronomen der Araber und ihre Werke. Abhandlungen zur Geschichte der mathematische Wissenschaft, 10. Leipzig, 1900. Reprint. New York: Johnson Reprint, 1972. Pp. ix + 277.

A list of Arabic scholars and their works; a number of these reached the Latin West in translation.

53. Talbot, Charles H. and E.A. Hammond. The Medical Practitioners of Medieval England: A Biographical Register. London: Wellcome Historical Medical Library, 1965. Pp. x + 503.

Supplies information on medical professionals in the British Isles, excluding Ireland, from the Anglo-Saxon period through the early sixteenth century.

54. Thorndike, Lynn and Pearl Kibre. A Catalogue of Incipits of Mediaeval Scientific Writings in Latin. Rev. and aug. ed. Cambridge, Mass.: Mediaeval Academy of America, 1963. Pp. xxii + 1938.

55. Wickersheimer, Ernest. Dictionnaire biographique des médecins en France au moyen âge. 3 vols. Paris, 1936. New facsimile edition under the direction of Guy Beaujouan. Geneva: Droz, 1979.

A list of physicians and members of associated medical professions with biographical details.

56. Zimmermann, Albert. Verzeichnis ungedruckter Kommentare zur Metaphysik und Physik des Aristoteles aus der Zeit von etwa 1250-1350, I. Leiden: Brill, 1971--. Pp. viii + 301.

Lists commentaries, mostly anonymous, with their question titles.

57. Zinner, Ernst. Verzeichnis der astronomischen Handschriften des deutschen Kulturgebietes. Munich: C.H. Beck, 1925. Pp. 544.

## ENCYCLOPEDIC TRADITION


58. Adelard of Bath. *Des Adelard von Bath Traktat De eodem et diverso*. Edited by Hans Willner. *Beiträge zur Geschichte der Philosophie des Mittelalters*, 4, 1. Münster: Aschendorff, 1903. Pp. viii + 112.

59. Adelard of Bath. *Dodi Venechdi (Uncle and Nephew), the Work of Berachya Hanakdan, now edited from MSS at Munich and Oxford ... to which is added the first English translation from the Latin of Adelard of Bath's Quaestiones naturales*. Edited by Hermann Gollancz. London: Oxford University Press, 1920.

    Adelard's *Natural Questions* on pp. 114-121.

60. Adelard of Bath. *Die Quaestiones naturales des Adelardus von Bath*. Edited by Martin Müller. *Beiträge zur Geschichte der Philosophie und Theologie des Mittelalters*, 31, 2. Münster: Aschendorff, 1934. Pp. 91.

61. Adelard of Bath. *Quaestiones naturales di Adelardo di Bath*. Edited by S. Boassi, A. di Giovanni and B. Ferrari. Rapallo: Canessa, 1965. Pp. 205.

62. Aiken, Pauline. "The Animal History of Albertus Magnus and Thomas of Cantimpré." *Speculum*, 22 (1947), 205-225.

    Argues by comparisons of passages in their works that Albert's *De animalibus* borrowed extensively from the encyclopedia of Thomas of Cantimpré.

63. Bartholomaeus Anglicus. *De rerum proprietatibus*. Frankfort, 1601. Reprint. Frankfort: Minerva, 1964. Pp. 1261.

64. Bate, Henry (of Malines). Speculum divinorum et quorundam naturalium. 2 vols. Edited by E. van de Vyver. Louvain: Publications universitaires, 1960, 1967.

65. Birkenmajer, Aleksander. "Henri Bate de Malines: Astronome et philosophe du XIIIe siècle." Études d'histoire des sciences et la philosophie du moyen âge. Edited by Aleksandra Maria Birkenmajer and Jerzy B. Korolec. Studia Copernicana, 1. Warsaw: Polish Academy of Sciences,1970, pp. 103-115.

Discusses Bate's life and work, especially his astrological-astronomical treatises. Focuses on a description of his compilation, the Speculum divinorum et quorundam naturalium.

66. Brehaut, Ernest. An Encyclopedist of the Dark Ages: Isidore of Seville. Studies in History, Economics and Public Law, 48. New York: Columbia University Press, 1912. Pp. 274.

Contains translations of extracts from most of the 20 books of Isidore's Etymologies as well as a complete translation of the section on geometry in Book III.

67. Cassiodorus. Cassiodori Senatoris Institutiones. Edited by R.A.B. Mynors. Oxford: Clarendon Press, 1937. Pp. lvi + 193.

68. Cassiodorus. An Introduction to Divine and Human Readings. Translated with an introduction and notes by L.W. Jones. New York: Columbia University Press, 1946. Pp. xvii + 233.

69. Engbring, Gertrude. "Saint Hildegard, Twelfth-Century Physician." Bulletin of the History of Medicine, 8 (1940), 770-784.

Discusses the content of two medical writings attributed to this twelfth-century mystic and visionary; their authenticity has, however, been rejected by Singer. The content of these works, practical and based on folk medicine, places them within the monastic medical tradition.

70. von Erhardt-Siebold, Erika and Rudolf von Erhardt. *The Astronomy of Johannes Scotus Erigena*. Baltimore: Williams & Wilkins, 1940. Pp. 69.

Argues based on a new translation and interpretation of the relevant passages of the *De divisione naturae* that Erigena's astronomy did not anticipate the Tychonian system. Indeed, there is nothing heliocentric in Erigena's cosmology, which was derived from Pliny.

71. von Erhardt-Siebold, Erika and Rudolf von Erhardt. *Cosmology in the Annotationes in Marcianum: More Light on Erigena's Astronomy*. Baltimore: Williams & Wilkins, 1940. Pp. 45.

Argues that the astronomy of Erigena's commentary on Martianus Capella is no more heliocentric than that of the *De divisione naturae*. See item 70.

72. Erigena, John Scotus. *Joannis Scotti Eriugenae Periphyseon (De divisione naturae)*, I & II. 2 vols. Edited by I.P. Sheldon-Williams with the collaboration of Ludwig Bieler. Dublin: Dublin Institute for Advanced Studies, 1968, 1972. Pp. x + 269, vi + 252.

73. Heyse, Elisabeth. *Hrabanus Maurus' Enzyklopadie De rerum naturis. Untersuchung zu den Quellen und zur Methode der Kompilation*. Munich: Arbeo-Gesellschaft, 1969. Pp. 163.

Concludes by an analysis of the sources, that Rabanus' work should be considered a continuation of Isidore's *Etymologies*.

74. Hildegard of Bingen. *Heilkunde: Das Buch von dem Grund und Wesen und der Heilung der Krankheiten*. Edited by Heinrich Schipperges. Salzburg: Otto Müller, 1957. Pp. 332.

74a. Hildegard of Bingen. *Welt und Mensch: Das Buch De operatione dei aus dem Genter Kodex*. Translated with a commentary by Heinrich Schipperges. Salzburg: Otto Müller, 1965. Pp. 358.

75. Honorius of Autun. *Honorius Augustodunensis Clavis physicae*. Edited by Paolo Lucentini. Rome: Edizioni di storia e letteratura, 1974. Pp. lix + 326.

An edition of the Latin text of the first part of the *Clavis*. Maintains that the work is partly a compendium and partly a reproduction of the *De divisione naturae* of John Scotus Erigena.

76. Hünemörder, Christian. "Die Bedeutung and Arbeitsweise des Thomas von Cantimpré und sein Beitrag zur Naturkunde des Mittelalters." *Medizinhistorisches Journal*, 3 (1968), 345-357.

Discusses the contents and sources of the *De rerum natura* and the differences between the three recensions of the work. Thomas' encyclopedia was one of the all-encompassing works of this kind written after the translation of Aristotle's zoological books.

77. Isidore of Seville. *Isidori Hispalensis Episcopi Etymologiarum sive originum libri XX*. 2 vols. Edited by W.M. Lindsay. Oxford: Clarendon Press, 1911.

78. Isidore of Seville. *The Medical Writings*. Translated with introduction and commentary by William D. Sharpe. *Transactions of the American Philosophical Society*, n. s., 54, part 2. Philadelphia: American Philosophical Society, 1964. Pp. 75.

Contains the anatomical and medical sections of the *Etymologies* including material from Book XI, "Man and Monsters," and Book IV, "On Medicine."

79. Isidore of Seville. *De natura rerum*. Edited by Gustavus Becker. Berlin, 1857. Reprint. Chicago: Argonaut, 1967. Pp. 87.

80. Lusignan, Serge. *Préface au Speculum maius de Vincent de Beauvais: Réfraction et diffraction*. *Cahiers d'études médiévales*, V. Paris: J. Vrin, 1979, Pp. 146.

An analysis and edition of the text of the *Libellus totius operis apologeticus* which serves as a preface to to the encyclopedic *Speculum maius* and in which Vincent sketched the rationale for his work.

81. Lusignan, Serge. "Les arts méchaniques dans le *Speculum doctrinale* de Vincent de Beauvais." *Cahiers d'études médiévales*, VII. *Les arts méchaniques au moyen âge*.

Edited by Guy H. Allard and Serge Lusignan. Paris: J. Vrin, 1982, pp. 33-48.

Discusses Vincent's treatment of the mechanic arts which retained the classification scheme of Hugh of St. Victor, but amplified treatment of these arts by extracts from authorities. In Vincent's list of the mechanic arts, however, medicine is replaced by alchemy.

82. Lutz, Cora E., ed. Johannis Scotti Annotationes in Marcianum. Cambridge, Mass.: Mediaeval Academy of America, 1942. Pp. xxx + 244.

83. Lutz, Cora E. "The Commentary of Remigius of Auxerre on Martianus Capella." Mediaeval Studies, 19 (1957), 137-156.

Argues that one of the two versions extant of the last four books of Remigius' commentary is indeed by him and not his more brilliant master, Scotus Erigena. The second of these versions may be by one of Remigius' pupils.

84. Macrobius. Commentary on the Dream of Scipio. Translated with introduction and notes by William H. Stahl. New York: Columbia University Press, 1952. Pp. xi + 278.

85. Neckham, Alexander. Alexandri Neckham De naturis rerum libri duo. Edited by Thomas Wright. Rerum Britannicarum medii aevi scriptores, 34. London: Green, Longman, Roberts and Green, 1863. Pp. lxxviii + 354.

86. Plato. Timaeus a Calcidio translatus commentarioque instructus. Edited by J.H. Waszink. Leiden: Brill, 1962. Pp. 436.

87. Quentin, Albrecht. Naturkenntnisse und Naturanschauungen bei Wilhelm von Auvergne. Hildesheim: Gerstenberg, 1976. Pp. xiv + 167.

Argues that William attempted to counteract the independence of nature and the limitations to God's power introduced into scholastic thought by Aristotle and his Arabic commentators.

88. Singer, Charles. "The Scientific Views and Visions of Saint Hildegard of Bingen." Studies on the History

<u>and Method of Science</u>. 2d ed. London: William Dawson and Sons, 1955, pp. 1-55.

Summarizes the difficulties in extracting the scientific views of this ecstatic visionary, in distinguishing her genuine works and in following her changes in viewpoint from earlier to later treatises. Hildegard wrote too early to be much influenced by the twelfth-century influx of scientific materials into the Latin West. Her cosmological views reflect the influence of Biblical sources, Augustine and Bernard Silvestris. Outside of contemporary anatomical diagrams, her anatomy and physiology had few sources on which to draw.

89. Stahl, William H. "Dominant Traditions in Early Medieval Latin Science." <u>Isis</u>, 50 (1959), 95-124.

Describes the derivative and superficial nature of the work of early medieval handbook compilers, commentators and encyclopedists who both plundered and concealed their sources, and had little real knowledge of technical Hellenistic science. See item 17.

90. Stahl, William H. "The <u>Quadrivium</u> of Martianus Capella: Its Place in the Intellectual History of Western Europe." <u>Arts libéraux et philosophie au moyen âge. Actes du quatrième congrès international de philosophie médiévale</u>. Montreal: Institut d'études médiévales, 1969, pp. 959-967.

Sketches the sources and treatment of the quadrivial arts in Martianus, a major and influential figure in the study of the seven liberal arts in the Latin West. See item 91.

91. Stahl, William H. <u>Martianus Capella and the Seven Liberal Arts</u>. 2 vols. New York: Columbia University Press, 1971, 1977.

Vol. I contains an analysis of Martianus' sources for the <u>quadrivium</u> and of the influence of the <u>Marriage</u> on medieval authors, with a study of the allegory and the <u>trivium</u> by Richard Johnson and E.L. Burge. Vol. II has a translation of the <u>Marriage</u> by Stahl, Johnson and Burge.

92. Thomas of Cantimpré. <u>Liber de natura rerum</u>. Vol. I. Berlin: de Gruyter, 1973--.

93. Thomas of Cantimpré. "De naturis rerum. Prologue, Book III and Book XIX." Edited by John Block Friedman. La science de la nature: Théories et pratiques. Cahiers d'études médiévales, II. Paris: J. Vrin, 1974, pp. 107-154.

Provides an edition of three portions of the twenty-volume De naturis rerum by the Flemish Dominican (b. 1201).

94. Thorndike, Lynn. "An Unidentified Work by Giovanni da' Fontana: Liber de omnibus rebus naturalibus." Isis, 15 (1931), 31-46.

Identifies this work, a blend of geography, astronomy, superstition and incredible stories as by the fifteenth-century Venetian author usually associated with works on military and hydraulic engineering. See item 1407.

95. Vincent of Beauvais. Speculum quadruplex sive speculum maius. 4 vols. Belleri, 1642. Reprint. Graz: Akademische Druck-u. Verlagsanstalt, 1964-65.

Contains the Speculum naturale, Speculum doctrinale, Speculum morale and Speculum historiale.

## TRANSMISSION AND TRANSLATION

96. Allan, Donald J. "Mediaeval Versions of Aristotle, *De caelo*, and of the Commentary of Simplicius." *Mediaeval and Renaissance Studies*, 2 (1950), 82-120.

    Argues that prior to and independent of the translations of William of Moerbeke, Grosseteste translated the *De caelo* to Book III, 1, and also Simplicius' commentary on the *De caelo* to this point, perhaps from Book I.

97. Alonzo Alonzo, Manuel. "Traducciones del arcediano Domingo Gundisalvo." *al-Andalus*, 12 (1947), 295-338.

    Discusses eleven philosophical works, giving MSS and editions of Latin versions, in which Dominicus Gundisalvus appears as chief translator, sometimes aided by his collaborator, Ibn Daūd, made between 1140-1186.

98. Alonso Alonso, Manuel. "Juan Sevillano, sus obras propias y sus traducciones." *al-Andalus*, 18 (1953), 17-49.

    Argues that John of Spain and John of Seville, while active in twelfth-century Spain, are two distinct persons. Lists the translations and original works of John of Seville (mostly astronomy and astrology) and the MSS where they can be found. See item 160.

99. d'Alverny, Marie-Thérèse. "Avendauth?" *Homenaje a Millás Vallicrosa*, I. Barcelona: Consejo superior de de investigaciones scientíficas, 1954, pp. 19-43.

Maintains that Avendauth should be identified as the Jewish scholar, Abraham ibn Daūd who worked as a translator in collaboration with Dominicus Gundissalinus. He should not be confused with John of Spain or John of Seville or any twelfth-century translator named John.

100. d'Alverny, Marie-Thérèse. "Les traductions d'Aristote et de ses commentateurs." XIIe congrès international d'histoire des sciences. Revue de synthèse, 89 (1968), 125-152.

Summarizes and lists the projects underway and work done (as of 1968) in preparing editions of Aristotelian and pseudo-Aristotelian works and commentaries extant in medieval translation from Greek and Arabic, as well as related Greek and Arabic philosophical works.

101. d'Alverny, Marie Thérèse. "Avicennisme en Italie." Oriente e occidente nel medioevo: Filosofia e scienze. Rome: Accademia nazionale dei Lincei, 1971, pp. 117-140.

Argues that knowledge of Avicenna's philosophical encyclopedia, the al-Shifā, and of his Canon was brought to Italy from Toledo by itinerant Italian translators and scholars. From the mid-thirteenth century, medical schools in Italy, especially Bologna, gave a prominent place to the Canon while, as arts and medicine were one faculty, medical commentators turned to the al-Shifā for explanations of Avicenna's medical work.

102. Beaujouan, Guy. "Fautes et obscuritiés dans des traductions médicales du moyen âge." XIIe congrès international d'histoire des sciences. Revue de synthèse, 89 (1968), 125-152.

Discusses the types of problems and sources of error arising from the specialized terminology in the medieval translations of medical texts from Arabic and Greek. Technical terms were often victims of misunderstanding and error which led to the introduction of peculiar and incomprehensible terms, often perpetuated and multiplied by bad reading of the Latin text by copyists.

103. Birkenmajer, Aleksander. "Eine wiedergefundene Übersetzung Gerhards von Cremona." Études d'histoire des sciences et la philosophie du moyen âge. Edited

by Aleksandra Maria Birkenmajer and Jerzy B. Korolec. Studia Copernicana, 1. Warsaw: Polish Academy of Sciences, 1970, pp. 22-31.

Reports a Gerard translation of a treatise by Alfarabi which surveys the content of the second half of the *Physics*. Includes the Latin text.

104. Birkenmajer, Aleksander. "Prolegomena in Aristotelem latinum: Classement des ouvrages attribués à Aristote par le moyen âge latin." *Études d'histoire des sciences et la philosophie du moyen âge*. Edited by Aleksandra Maria Birkenmajer and Jerzy B. Korolec. Studia Copernicana, 1. Warsaw: Polish Academy of Sciences, 1970, pp. 55-71.

A classification of the Latin works which circulated from the twelfth to the fifteenth centuries under Aristotle's name.

105. Birkenmajer, Aleksander. "Le rôle joué par les médecins et les naturalistes dans la réception d'Aristote au XIIe et XIIIe siècles." *Etudes d'histoire des sciences et la philosophie du moyen âge*. Edited by Aleksandra Maria Birkenmajer and Jerzy B. Korolec. Studia Copernicana, 1. Warsaw: Polish Academy of Sciences, 1970, pp. 73-87.

Argues that the introduction and first absorption of Aristotle in the Latin West before c. 1200-1230 was due to the efforts of scholars interested in natural science and medicine. Shortly after 1230, the new Aristotle was increasingly utilized by theologians, but the way had been opened for this by their naturalist and medical precursors.

106. Björnbo, A.A. "Hermannus Dalmata als Übersetzer astronomischer Arbeiten." *Biblioteca mathematica*, ser. 3, 4 (1903), 130-133.

Discusses Hermann's role in the translation of Ptolemy's *Planisphere* and of Albumasar's *Introductorium*.

107. Bonora, Fausto and George Kern. "Le traduzioni greco-latine di opere mediche del basso medioevo e la lora importanza nella valutazione della cultura pre-unamistica." *Rivista di storia della medicina*, 21 (1977), 210-218.

Argues that the classical tradition continued in ecclesiastical centers in Italy during the early Middle Ages. Likewise there were connections between the churches of Rome and Byzantium as well as commercial, political and military contacts between the East and the Latin West. Focuses on Burgundio of Pisa who translated works in theology, law and agriculture, but particularly the medical works of Galen and Hippocrates.

108. Busard, H.L.L. "Über die Überlieferung der *Elemente* Euklids über die Länder des nahen Ostens nach West-Europa." *Historia mathematica*, 3 (1976), 279-290.

Discusses the transmission of the Arabic tradition of the *Elements* to the Latin West in Adelard I and II, and in the versions of Hermann of Carinthia and Gerard of Cremona. The Arabic tradition comprised two versions by al-Ḥaggāg (I and II) and that due to two Arabic scholars, called for short the Isḥāq-Tābit version. Argues that Adelard I and Hermann of Carinthia rely on al-Ḥaggāg I; Adelard II on al-Ḥaggāg II. On the other hand, Gerard followed Isḥāq-Tābit although his translation has proofs from al-Ḥaggāg I. Discusses also the relation between the Arabic-Latin tradition and the Greek edition of Heiberg.

109. Clagett, Marshall. "The Medieval Latin Translations from the Arabic of the *Elements* of Euclid, with Special Emphasis on the Versions of Adelard of Bath." *Isis*, 44 (1953), 16-42.

Maintains that there were three principal versions of the *Elements* made from Arabic in the twelfth century by Adelard of Bath: a translation, an abridgment and a paraphrase or edition. Other translations from Arabic were made by Hermann of Carinthia and Gerard of Cremona.

110. Clagett, Marshall. "King Alfred and the *Elements* of Euclid." *Isis*, 45 (1954), 269-277.

Determines from MS evidence that the attribution of a translation or commentary on Euclid to Alfred as early as the tenth century is a legend.

111. Clagett, Marshall. "Archimedes in the Late Middle Ages." Perspectives in the History of Science and Technology. Edited by Duane H.D. Roller. Norman: University of Oklahoma Press, 1971, pp. 239-259.

Summarizes the transmission of the Archimedean corpus to the Arabs and the Latin West and the use made of these treatises. Collected by Byzantine scholars at Constantinople from the sixth to the ninth centuries, many works found their way to the Arabic world. In the twelfth century translations from Arabic appear in the West made by Gerard of Cremona and Plato of Tivoli followed by the 1269 translations of the Byzantine corpus by Moerbeke. These materials were used by scholastics at Paris; a new translation was made by Jacobus Cremonensis about 1450.

112. Días y Días, Manuel C. "Les arts libéraux d'après les écrivains espagnols et insulaires aux VIIe et VIIIe siècles." Arts libéraux et philosophie au moyen âge. Actes du quatrième congrès international de philosophie médiévale. Montreal: Institut d'études médiévales, 1969, pp. 37-46.

Points out developments in the early Middle Ages in the transmission of ancient thought and the classification of the sciences as seen in Spain (largely in the works of Isidore of Seville), in Ireland and Anglo-Saxon England.

113. Dunlop, D.M. Arabic Science in the West. Karachi: Pakistan Historical Society, 1958? Pp. v + 119.

Traces paths of transmission of Arabic science to the Christian West beginning in tenth-century Spain, and the use of these materials by Michael Scot, Albertus Magnus and Peter of Abano.

114. Dunlop, D.M. "The Work of Translation at Toledo." Babel, 6 (1960), 55-59.

Summarizes the activity of the twelfth-century translation school at Toledo, and discusses the work of Millás Vallicrosa, Alonzo Alonzo and d'Alverny in identifying important translators and authors.

115. Goldstein, Bernard R. "The Role of Science in the Jewish Community in Fourteenth-Century France." Machaut's World: Science and Art in the Fourteenth Century. Edited by Madeleine Pelner Cosman and Bruce Chandler. New York: New York Academy of Sciences, 1978, pp. 39-49.

Shows that the Jewish community in fourteenth-century Provence, originally from Moslem Spain, contained many members who not only translated scientific and philosophical works into Hebrew, but made original contributions in mathematics and astronomy. Levi ben Gerson was the most original among this group of Hebrew writers, and some of his astronomy was translated into Latin. He is best known for his invention of the Jacob Staff, an instrument widely used in navigation.

116. Grabmann, Martin. Forschung über die lateinischen Aristoteles: Übersetzungen des XIII. Jahrhunderts. Beiträge zur Geschichte der Philosophie des Mittelalters, 17, 5-6. Münster: Aschendorff, 1916. Pp. xxvii + 270.

Discusses the use of Aristotelian materials in the works of thirteenth-century schoolmen, provides MS traditions and lists translators of individual works.

117. Grabmann, Martin. Mittelalterliche lateinische Aristotelesübersetzungen und Aristoteleskommentare in Handschriften spanischer Bibliotheken. Sitzungsberichte der Bayerischen Akademie der Wissenschaften, philos.-philolog. und hist. Klasse, 5. Munich, 1928. Pp. 120.

Describes MSS of translations from Greek and Arabic of various Aristotelian works and of commentaries on Aristotle in Spanish collections.

118. Grabmann, Martin. Mittelalterliche lateinische Übersetzungen von Schriften der Aristoteles-Kommentatoren Johannes Philoponos, Alexander von Aphrodisias und Themistios. Sitzungsberichte der Bayerischen Akademie der Wissenschaften, phil.-hist. Klasse, 7. Munich, 1929. Pp. 72.

Discusses the MSS and translators of Philoponus' commentary to Book III of the De anima, Alexander's De fato ad imperatores and Themistius' commentary on the Prior analytics and the De anima.

119. Grabmann, Martin. Guglielmo di Moerbeke, O. P., il traduttore delle opere di Aristotele. Rome: Pontificia università à Gregoriana, 1970. Pp. xi + 193.

First published in 1946. Discusses the life and career of the Flemish Dominican and his association with Campanus of Novara, Witelo and Aquinas as well as his translations of Aristotle, the Greek commentators, Archimedes and others.

120. Grabmann, Martin. "Kaiser Friedrich II. und sein Verhältnis zu aristotelischen und arabischen Philosophie." Mittelalterliches Geistesleben, II. Munich, 1936. Reprint. Hildesheim: Olms, 1975, pp. 103-137.

Maintains that Frederick's enormous interest in Aristotelian and Arabic natural philosophy is evidenced by his encouragement of translators and commentators at the court at Palermo and at the University of Naples, as well as by his own De arte venandi and the questions on scientific subjects he put to Arabic scholars (the Sicilian questions).

121. Grant, Edward. "Henricus Aristippus, William of Moerbeke and Two Alleged Mediaeval Translations of Hero's Pneumatica." Speculum, 46 (1971), 656-669.

Concludes by new evidence and a reinterpretation of previous claims, that neither Aristippus nor Moerbeke translated the Pneumatica. This accounts for the lack of influence of Hero's work on medieval treatments of the vacuum and the lack of Latin MSS of the treatise until near the end of the fifteenth century.

122. Haskins, Charles H. Studies in the History of Mediaeval Science. 2d ed. Cambridge, Mass., 1927. Reprint. New York: Ungar, 1960. Pp. xx + 411.

A pioneering study of the twelfth- and thirteenth-century translators from Arabic and Greek prominent in the scientific revival of the Latin West. Also includes material on twelfth-century astronomical writers, the introduction of Arabic science into England and scientific activities at the court of Frederick II. Although details have been augmented by subsequent research, the study remains valuable.

123. Hossfeld, Paul. "Der Gebrauch der Aristotelischen Übersetzung in den Meteora des Albertus Magnus." Mediaeval Studies, 42 (1980), 395-406.

Discusses the translations used by Albert in his commentary on the Meteorology. Of the work's four books, Albert used a translation from Arabic for the first three and a translation from Greek for the fourth book. It is difficult to separate Albert's own opinion from the text itself.

124. Hourani, George F. "The Medieval Translations from Arabic to Latin Made in Spain." The Muslim World, 62 (1972), 97-114.

Traces the scientific culture of tenth- and eleventh-century Andalusia and its diffusion to Latin Europe through the efforts of Muslims, Jews and Christians from the beginning of the eleventh century. Discusses the types of works translated and their influence and lists the chief Latin translations from Greek and Arabic of philosophical and scientific works from 1100 to 1328.

125. Humbert, Pierre. "L'astronomie au XIIIe siècle dans la France méditerranéene." Revue des questions scientifiques, 28 (1935), 204-215.

Discusses the role played by centers of learning in Provence and the Languedoc in the absorption and transmission of Arabic astronomical materials.

126. Kennedy, E.S. "Planetary Theory in the Medieval Near East and its Transmission to Europe." Oriente e occidente nel medioevo: Filosofia e scienze. Rome: Accademia nazionale dei Lincei, 1971, pp. 595-604.

Discusses various mathematical devices and techniques employed by eastern astronomers to describe the motions of the planets, sun and moon. Some of this material reached the Latin West, but was generally received uncritically. The West did have the work of al-Battānī which included all of Ptolemaic astronomy, and the homocentric system devised by al-Biṭrūgī which could not compete with Ptolemy. Eastern astronomers also invented non-Ptolemaic uniform circular motion models, but these were unknown in the West before Copernicus.

126a. Kurdziałek, Marian. "David von Dinant als Ausleger der Aristotelischen Naturphilosophie." Die Auseinandersetzungen an der Pariser Universität im XIII. Jahrhundert. Edited by Albert Zimmermann. Berlin: de Gruyter, pp. 181-192.

Argues that David, familiar with Greek, was responsible for the early translation and transmission of part of the Aristotelian corpus which he also used in his own work and discusses the opinions which led to his condemnation on charges of pantheism. David's interpretation of his sources contributed to the prohibition of Aristotelian natural philosophy in the early thirteenth century.

127. Lattin, Harriet Pratt. "Lupitus Barchinonensis." Speculum, 7 (1932), 58-64.

Suggests that the Lupitus of Barcelona to whom Gerbert wrote requesting an astronomical work can be identified with the archdeacon Lobetus also called Seniofredus.

128. Lemay, Richard. "Fautes et contresens dans les traductions arabo-latines médiévales: L'Introductorium in astronomiam d'Abou Ma^c shar de Balkh." XIIe congrès international d'histoire des sciences. Revue de synthèse, 89 (1968), 101-123.

Studies the types of errors in the transmission of the Introductorium from Arabic to Latin by a three-fold comparison (errors of reading, of interpretation, textual or cultural contamination) in a critical edition of the translation by John of Seville, the translation of Hermann of Carinthia, the revision made by Gerard of Cremona and the Arabic text.

129. Marenghi, Gerardo. "Un capitolo dell'Aristotele medievale: Bartolomeo da Messina traduttore dei Problemata physica." Aevum, 36 (1962), 268-283.

Analyses the translation techniques of Bartholomew of Messina, thirteenth-century translator of Aristotelian and pseudo-Aristotelian works from Greek at the court of Manfred in Sicily, as employed in his translation of the Problemata.

130. Mercken, H. Paul. "Medieval Translations of Greek Philosophy: A Computer Study." Sprache und Erkenntnis im Mittelalter: Akten des VI. internationalen Kongresses für mittelalterliche Philosophie, II. Berlin: de Gruyter, 1981, pp. 684-692.

Describes a computer project designed to arrive at the translation procedures used by Grosseteste in his translation of part of the Nicomachean Ethics and material connected with this text.

131. Millás Vallicrosa, J.M. "Translations of Oriental Scientific Works (to the End of the Thirteenth Century)." The Evolution of Science. Edited by Guy S. Métraux and François Crouzet. New York: New American Library, 1963, pp. 128-167.

Traces translators from Arabic and their works from the first translations made at Santa Maria de Ripoll in Catalonia and at Salerno, translations made in the late eleventh and early twelfth centuries, those made by scholars who worked in the Pyrenees and the Ebro valley as well as those due to the Toledan School and the scholars at the court of Alfonso X.

132. Minio-Paluello, Lorenzo. "The Genuine Text of Boethius' Translation of Aristotle's Categories." Mediaeval and Renaissance Studies, 1 (1941-43), 151-177.

Argues that while Boethius did translate and comment on the Categories, the translation in many MSS and printed editions from the tenth century on is not due to Boethius, whose own translation has been found, but to a translator whose work is in part of the tenth century.

133. Minio-Paluello, Lorenzo. "Henri Aristippe, Guillaume de Moerbeke et les traductions latines médiévales des Météorologiques et du De generatione et corruptione d'Aristote." Revue philosophique de Louvain, 3d sér., 45 (1947), 206-235.

Argues that the De generatione et corrputione as we have it was not translated from Greek by Aristippus. Further, it is not possible to speak of three Latin versions of the De generatione, nor of a revision of the work by Moerbeke. Likewise, Moerbeke made a new translation of the Meteorology and did not simply revise the fourth book.

134. Minio-Paluello, Lorenzo. "Note sull'Aristotele latino medievale." Rivista di filosofia neo-scholastica, 42 (1950), 222-237.

Presents three notes concerning Aristotelian translations from Greek. Concludes: (1) The Metaphysica vetustissima contained all of the Metaphysics; (2) The Physica vaticana and the Metaphysica media were translated by the same person; (3) The two translations of the pseudo-Aristotelian De mundo can be attributed to Nicolas Siculo (Greco) and to Bartholomew of Messina.

135. Minio-Paluello, Lorenzo. "Iacobus Veneticus Grecus: Canonist and Translator of Aristotle." Traditio, 8 (1952), 265-304.

Traces the translations and influence of James the Venetian and the Greek who was a most successful pioneer in the translation of Aristotelian works from Greek to Latin in the twelfth century. He was probably the first to translate the Physics, De anima, the Metaphysics and parts of the Parva naturalia, and he retranslated several of Aristotle's logical treatises. Whether he was a Venetian who lived in Greece, or a Venetian by descent who was born and educated in the Greek world, or a Greek born in Venice is not known.

136. Muckle, J.T. "Greek Works Translated Directly into Latin before 1350." Mediaeval Studies, 4 (1942), 33-42.

Lists the translations and paraphrases of primarily theological and patristic works.

137. Murdoch, John E. "Euclides graeco-latinus: A Hitherto Unknown Medieval Latin Translation of the Elements Made Directly from the Greek." Harvard Studies in Classical Philology, 71 (1967), 249-302.

Reports that the long-held suspicion of the existence of a Latin version of the Elements done directly from Greek has been confirmed by the discovery of two MSS, one in the Bibliothèque Nationale at Paris containing Books I-XIII and abbreviated treatments of Books XIV and XV, the other at the National Library at Florence, which is incomplete. Suggests that the translator of this version of the Elements is identical with the translator of the twelfth-century version of the Almagest from Greek made in Sicily.

138. Murdoch, John E. "The Medieval Euclid: Salient Aspects of the Translations of the Elements by Adelard of Bath and Campanus of Novara." XIIe congrès international d'histoire des sciences. Revue de synthèse, 89 (1968), 67-94.

Discusses the nature and fortunes of the three twelfth-century versions of Euclid and the thirteenth-century adaptation made by Campanus of Novara. Compares the Adelardian tradition and the Greek Euclid as to missing propositions or definitions, changed order of propositions and misunderstanding of enunciations or definitions. Characterizes the Adelardian tradition, pointing out the additions of a pedagogical nature, the emphasis on logical structure and philosophical issues and the tension between geometry and arithmetic.

139. Murdoch, John E. "Transmission and Figuration: An Aspect of the Islamic Contribution to Mathematics, Science and Natural Philosophy in the Latin West." The Commemoration Volume of Bīrūnī International Congress in Tehran. B. English and French Papers. Tehran: High Council of Culture and Art, 1976, pp. 407-437.

Argues that Islam's contribution to the science of the Latin West lay more in the transmission of Greek materials by Arabic intermediaries than in the direct Latin translation of works of Arabic origin. Arabic scholars transformed the Greek science and philosophy which returned to the West. The point is illustrated by two examples: the Arabic-Latin Euclid and the transformation of Aristotelian natural philosophy in the commentaries of Averroes.

140. Opelt, Ilona. "Zur Übersetzungstechnik des Gerhards von Cremona." Glotta, 38 (1959), 135-170.

Investigates Gerard's translation technique revealed in his translation of the De caelo et mundo from

Arabic by a comparison of his work with the Arabic and Greek texts, pointing out the literal nature of the translation with its Arabic elements of phraseology, syntax and style.

141. Otte, James K. "The Life and Writings of Alfredus Anglicus." Viator, 3 (1972), 275-291.

Augments and summarizes previous knowledge concerning this author. Alfred, who apparently visited Spain, was a translator of scientific works from Arabic, a commentator on the new Aristotle and composed an independent treatise, the De motu cordis. However, it remains difficult to connect him with a specific center of learning in England.

142. Otte, James K. "The Role of Alfred of Sareshel (Alfredus Anglicus) and his commentary on the Metheora in the Reacquisition of Aristotle." Viator, 7 (1976), 197-209.

Focuses on Alfred's role as the first Latin commentator of Aristotle and his sources used in the commentary on the Meteorology. Alfred drew on Greek and Arabic materials, including Aristotle, Avicenna, Alfarabi and Albumasar.

143. Pelster, Franz. "Neuere Forschungen über Aristotelesübersetzungen des 12. and 13. Jahrhunderts: Eine kritische Übersicht." Gregorianum, 30 (1949), 46-77.

Summarizes the work of scholars as of 1949 in establishing the dates of Aristotelian translations, primarily from Greek.

144. Procter, Evelyn S. "The Scientific Works of the Court of Alfonso X of Castile: The King and his Collaborators." Modern Language Review, 40 (1945), 12-29.

Discusses the astronomical and astrological works translated into Castillian from Arabic sources at Alfonso's court. Points out that the king was not merely a patron of science and learning, but encouraged projects of cooperative scholarship in which he served as general editor. Identifies some of the Jews, Spaniards and Italians who collaborated in these projects.

145. Romano, Davide. "Le opere scientifiche di Alfonso X e l'intervento degli Ebrei." Oriente e occidente nel medioevo: Filosofia e scienze. Rome: Accademia nazionale dei Lincei, 1971, pp. 676-711.

Stresses the important role played by Jewish participants in the scientific translation activity under the patronage of Alfonso X in the second half of the thirteenth century. Discusses the known Jewish contributors and their works.

146. Romano, Davide. "La transmission des sciences arabes par les juifs en Languedoc." Juifs et judaïsme de Languedoc XIIIe siècle-début XIV siècle. Edited by Marie-Humbert Vicaire and Bernard Blumenkranz. Toulouse: Édouard Privat, 1977, pp. 363-386.

Focuses on the role played by the family ibn Tibbon, originally from Spain, who brought Arabic and Greek scientific materials to their co-religionists in southern France.

147. Russell, Josiah C. "Hereford and Arabic Science in England about 1175-1200." Isis, 18 (1932), 14-25.

Argues that the presence of prominent scholars and translators from Arabic such as Roger of Hereford, Alfred of Sareshel and Daniel of Morley (who may have studied there), made Hereford a center for the transmission of Arabic science to England in the second half of the twelfth century. There is some literary evidence that a cathedral school may have existed at Hereford.

148. Schipperges, Heinrich. "Die frühen Übersetzer der arabischen Medizin in chronologischer Sicht." Sudhoffs Archiv für Geschichte der Medizin und der Naturwissenschaften, 39 (1955), 53-93.

Provides a chronological list of early translators of Arabic medical works with short biographical notices, lists of treatises, MSS and editions, and comments on the significance of the works.

149. Schipperges, Heinrich. "Zur Rezeption und Assimilation arabischen Medizin im frühen Toledo." Sudhoffs Archiv für Geschichte der Medizin und der Naturwissenschaften, 39 (1955), 261-283.

Discusses the program of translation of Arabic works at Toledo, ca. 1130-1150, with information on materials translated, translators, sources, MSS and editions.

150. Schipperges, Heinrich. "Das griechisch-arabische Erbe Toledos und sein Auftrag für die abendländische Heilkunde." Sudhoffs Archiv für Geschichte der Medizin und der Naturwissenschaften, 41 (1957), 113-142.

Assesses the role of the translation school at Toledo, and the extent to which medical thought of the Latin West was a legacy of this activity.

151. Schipperges, Heinrich. Die Assimilation der arabischen Medizin durch das lateinische Mittelalter. Sudhoffs Archiv für Geschichte der Medizin und der Naturwissenschaften. Beiheft 3. Wiesbaden: Steiner, 1964. Pp. ix + 240.

Treats the reception of Greek-Arabic medicine in the Latin West as stemming from three sources: Salerno, the Arabic versions of Aristotle, and the translation activity at Toledo. Further surveys the assimilation of this new material in France (Chartres, Paris, southern France), in England and in southern Italy.

152. Silverstein, Theodore. "Hermann of Carinthia and Greek: A Problem in the 'New Science' of the 12th Century." Medioevo e rinascimento: Studi in onore di Bruno Nardi, II. Florence: Sansoni, 1955, pp. 681-699.

Disputes the claim of Alonzo Alonzo in his edition of Hermann's De essentiis that this translator from Arabic also translated works from Greek. Argues from these De essentiis passages that Hermann did not draw directly from any Greek text and that his Greek materials were accessible in other languages. See item 614.

153. Steinschneider, Moritz. Die europäischen Übersetzungen aus dem arabischen bis Mitte des 17. Jahrhunderts. Vienna, 1904-05. Reprint. Graz: Akademische Druck- u. Verlagsanstalt, 1956. Pp. xii + 84 + 108.

Lists works of known translators and translations of works of known authors whose translations are unknown or uncertain. Reprinted from Sitzungsberichte der kaiserlichen Akademie der Wissenschaften in Wien, phil.-hist. Klasse.

154. Sudhoff, Karl. "Die kurze 'Vita' und das Verzeichnis der Arbeiten Gerhards von Cremona, von seinen Schulern und Studiengenossen kurz nach dem Tode des Meisters (1187) zu Toledo verabfasst." Archiv für Geschichte der Medizin, 8 (1914-15), 73-82. Reprint. Wiesbaden: Steiner, 1964.

The text of the "life" of Gerard, with a list of the works he translated from Arabic, composed by his students and colleagues at Toledo.

155. Thompson, J.W. "The Introduction of Arabic Science into Lorraine in the Tenth Century." Isis, 12 (1929), 184-193.

Argues that traditional connections between Spain and Europe beyond the Pyrenees continued between Moslem Spain and the Carolingian and Ottonian empires, and initiated the transmission of Arabic science to the West. Due to the efforts of prominent monastic reformers, who had cultural and diplomatic contacts with Spain, Arabic science moved into Lotharingia before Gerbert.

156. Thorndike, Lynn. "Date of the Translation by Ermengand Blasius of the Work on the Quadrant by Profatius Judaeus." Isis, 26 (1937), 306-309.

Confirms the date 1290 for Ermengand's translation based on new MS evidence.

157. Thorndike, Lynn. "The Latin Translations of the Astrological Tracts of Abraham Avenezra." Isis, 35 (1944), 293-302.

Points out that there are four known Latin translators of the short tracts on astrology by this author. Lists the works, the translator where known, extant MS versions, and provides a concordance indicating what versions appear together and their order in representative MSS and the 1507 edition.

158. Thorndike, Lynn. "Translations of Works of Galen from the Greek by Niccolò da Reggio (c. 1308-1345)." Byzantina Metabyzantina, 1 (1946), 213-235.

Lists more than 50 Galenic works translated by Niccolò, combining information from previous investigations supplemented by new materials. Entries are listed alphabetically by title of work with editions, MSS and citations to previous sources on Niccolò's translation activity.

159. Thorndike, Lynn. "The Latin Translations of Astrological Works by Messahalla." Osiris, 12 (1956), 49-72.

Lists MSS of and discusses several treatises by or attributed to Messahalla. These include fragmentary astrological materials, texts on interrogations, nativities, properties of stars as well as the genuine Epistola on eclipses, conjunctions and revolutions of the planets.

160. Thorndike, Lynn. "John of Seville." Speculum, 34 (1959), 20-38.

Differentiates Avendauth (Ibn Daw̄d), the Jewish scholar who worked as a translator with Dominicus Gundisalvus, from John of Seville, who translated Arabic astronomical-astrological materials, and composed works on these subjects. See item 99.

161. Thorndike, Lynn. "The Three Latin Translations of the Pseudo-Hippocratic Tract on Astrological Medicine." Janus, 49 (1960), 104-129.

Presents a detailed comparison of three Latin versions (translated from Greek by Peter of Abano and William of Moerbeke plus an anonymous version possibly from Arabic) of a treatise on the influence of the moon in the 12 signs of the zodiac commonly known as the Astronomia or Astrologia Ypocrates, taken from texts in numerous MSS. Gives the text for two signs, Taurus and Pisces, from several MSS for each translation. See item 1321.

162. Vuillemin-Diem, Gudrun. "Jacob von Venedig und der Übersetzer der Physica Vaticana und Metaphysica media." Archives d'histoire doctrinale et littéraire du moyen âge, 41 (1974), 7-25.

Maintains that the Physica vetus, the translation by James of Venice,is independent of the Physica

Vaticana. Likewise, James' Metaphysics translation (Metaphysica vetustissima) is independent of the Metaphysica media.

163. van de Vyver, A. "Les premières traductions latines (Xe-XIe siècles) de traités arabes sur l'astrolabe." Ier congrès international de geographie historique, 2 (1930), 266-290.

Traces the interrelationships of the MSS of the corpus of early astrolabe treatises including the source of Hermannus Contractus' De mensura astrolabi.

164. van der Vyver, A. "Les plus anciens traductions latines médiévales (Xe-XIe siècles) de traités d'astronomie et d'astrologie." Osiris, 1 (1936), 658-691.

Lists early works on astronomy and astrology, chiefly the latter, giving MS references, relations with other authors, citations by modern scholars.

165. Welborn, M.C. "Lotharingia as a Center of Arabic and Scientific Influence in the Eleventh Century." Isis, 16 (1931), 188-199.

Argues that the schools of Lorraine were especially devoted to science toward the end of the tenth century. Lotharingian geometers and astronomers were familiar with the astrolabe and scholars from Lorraine were imported to England; one of them, Walcher of Malvern, brought the astrolabe to England in the eleventh century.

166. Wingate, S.D. The Mediaeval Latin Versions of the Aristotelian Scientific Corpus, with Special Reference to the Biological Works. London: Courier Press, 1931. Pp. 136.

Discusses the paths by which Aristotle's scientific works entered the Latin West. Presented in chronological order from the translations of Boethius through humanistic versions, the discussion includes transmissions from Greek and Arabic.

167. Wüstenfeld, F. Die Übersetzungen arabischer Werke in das Lateinische seit dem XI. Jahrhundert. Abhandlungen der königlichen Gesellschaft der Wissenschaften zu Göttingen, 22. Göttingen, 1877. Pp. 133.

## CLASSIFICATION OF THE SCIENCES

168. Alfarabi (al-Fārābī). Über den Ursprung der Wissenschaften (De ortu scientiarum). Edited by Clemens Baeumker. Beiträge zur Geschichte der Philosophie des Mittelalters, 19. Münster: Aschendorff, 1916. Pp. 32.

169. Alfarabi (al-Fārābī). Catálogo de las sciencias. 2d ed. Edited by A. González Palencia. Madrid: Consejo superior de investigaciones científicas, 1953. Pp. xix + 176.

    Provides the text in the version of Gerard of Cremona with Spanish translation, and the text of the edition by Camerarius, printed in 1938, of the Latin version by Dominicus Gundisalvus.

170. Aquinas, Thomas. The Division and Methods of the Sciences. Questions V and VI of his Commentary on the De trinitate of Boethius. 3d rev. ed. Translated with introduction and notes by Armand Maurer. Toronto: Pontifical Institute of Mediaeval Studies, 1963. Pp. xl + 104.

    Points out that these sections of Aquinas' work contain the most extensive treatment of his division of the speculative sciences and their methods of procedure. The scheme is based on Aristotle-Boethius, but has additions and emendations.

171. Federici Vescovini, Graziella. "La 'perspectiva' nell' enciclopedia del sapere medievale." Vivarium, 6 (1968), 35-45.

    Traces the development of the place of optics within the classification of the theoretical sciences and among the quadrivial disciplines. Optics was first included in the *quadrivium* by Dominicus Gundisalvus, following Alkindi. Dominicus de Clavasio, in a question on the status of optics as a science and its position in relation to the mathematical disciplines, made it a member of the *quadrivium* and emphasized its role as a *scientia media*, between mathematics and physics.

172. Federici Vescovini, Graziella. "L'inserimento della *perspectiva* tra le arti del quadrivio." *Arts libéraux et philosophie au moyen âge*. *Actes du quatrième congrès international de philosophie médiévale*. Montreal: Institut d'études médiévales, 1969, pp. 969-974.

Maintains that in his *Quaestiones super perspectiva*, Dominicus de Clavasio added *perspectiva* or optics to the traditional *quadrivium*.

173. Frankowska-Teslecka, Małgorzata. "*Scientia* as Conceived by Roger Bacon." *Organon* (Warsaw), 6 (1969), 209-231.

Argues that Bacon's division of the sciences went beyond its Aristotelian and Arabic sources in its emphasis on languages, and in its focus on the interrelationship of all the sciences and on the connection between their speculative and practical aspects. Likewise Bacon's emphasis on mathematics and experimentation as essential to scientific method mark a creative continuation of Grosseteste.

174. Gagné, Jean. "Du *quadrivium* aux *scientiae mediae*." *Arts libéraux et philosophie au moyen âge*. *Actes du quatrième congrès international de philosophie médiévale*. Montreal: Institut d'études médiévales, 1969, pp. 975-986.

Traces views of the place of the intermediate sciences (*scientiae mediae*)--astronomy, music and optics--which focused on whether these sciences pertained more to mathematics or natural philosophy.

175. Gundisalvo, Domingo. *De divisione philosophiae*. Edited by Ludwig Baur. *Beiträge zur Geschichte der Philosophie des Mittelalters*, 4, 2-3. Münster: Aschendorff, 1903. Pp. xii + 408.

176. Hugh of St. Victor. Didascalicon. Edited by C.H. Buttimer. Washington, D.C.: Catholic University Press, 1939. Pp. lii + 160.

177. Hugh of St. Victor. Didascalicon: A Medieval Guide to the Arts. Translated with introduction and notes by Jerome Taylor. New York: Columbia University Press, 1961. Pp. xii + 254.

178. Kilwardby, Robert. De ortu scientiarum. Edited by Albert G. Judy. Toronto: Pontifical Institute of Mediaeval Studies, 1976. Pp. lxi + 255.

179. Lutz, Cora E. "Remigius' Ideas on the Classification of the Seven Liberal Arts." Traditio, 12 (1956), 65-86.

Maintains that the pedagogue, Remigius of Auxerre, in his commentary on Martianus, served to insure the preservation of the seven liberal arts within a context where the goal of knowledge was divine wisdom.

180. McKeon, Richard P. "The Organization of Sciences and the Relations of Cultures in the Twelfth and Thirteenth Centuries." The Cultural Context of Medieval Learning. Edited by John E. Murdoch and Edith D. Sylla. Dordrecht: Reidel, 1975, pp. 151-192.

Argues that cultures may be interpreted as structures of values which during the medieval period can be revealed in the organization of the sciences in the Greek, Latin, Hebrew and Moslem traditions.

181. Muhsin, Madhi. "Science, Philosophy, and Religion in Alfarabi's Ennumeration of the Sciences." The Cultural Context of Medieval Learning. Edited by John E. Murdoch and Edith D. Sylla. Dordrecht: Reidel, 1975, pp. 113-147.

Describes and discusses the content of five chapters in Alfarabi's classification of the sciences.

182. Sharp, D.E. "The De ortu scientiarum of Robert Kilwardby (d. 1279)." The New Scholasticism, 8 (1934), 1-30.

Sketches the content of Kilwardby's classification of the sciences arguing that it is based essentially on the Aristotelian-Boethian scheme and reveals little influence of Gundissalinus.

183. Steneck, Nicholas H. "A Late Medieval _Arbor scientiarum_." _Speculum_, 50 (1975), 245-269.

Analyzes the fourteenth-century _arbor scientiarum_, or classification of the sciences, in the _Expositio prologi Biblie_ of Henry of Hesse, written for his theology lectures at Vienna. Although dependent primarily on Augustinian sources and expressing traditional views, Henry's scheme is original in conception. The roots of the tree are the liberal arts whose importance Henry stresses. In his view the fruits of the tree are those arts which serve human needs or have social utility and should be cultivated.

184. Vermeirre, André. "La navigation d'après Hugues de Saint-Victor et d'après la pratique au XIe siècle." _Cahiers d'études médiévales_, VII. _Les arts mécaniques au moyen âge_. Edited by Guy H. Allard and Serge Lusignan. Paris: J. Vrin, 1982, pp. 51-61.

Points out the attention given to navigation in Hugh of St. Victor's treatment of the mechanic arts in his _Didascalicon_ and the approbation he extended to the profession of merchant. His attitude was colored by the milieu in which he lived and taught.

185. Weisheipl, James A. "Classification of the Sciences in Medieval Thought." _Mediaeval Studies_, 27 (1965), 54-90.

Traces the development of classification from the schemes introduced by Boethius, Augustine, Cassiodorus and Isidore, a Latin tradition culminating in Hugo of St. Victor, through the impact of Arabic and Neoplatonic notions stemming from Alfarabi which influenced the classification treatises of Dominicus Gundisalvus and Kilwardby. However, both Albertus Magnus and Aquinas took issue with the classification schemes in those treatises with Neoplatonic emphases.

# PHYSICAL SCIENCE AND NATURAL PHILOSOPHY

## GENERAL

186. Bacon, Roger. *The Opus maius of Roger Bacon*. 3 vols. Edited by J.H. Bridges. Oxford, 1897-1900. Reprint. Frankfort: Minerva, 1964.

187. Bacon, Roger. *Opus maius*. 2 vols. Translated by Robert B. Burke. Philadelphia, 1928. Reprint. New York: Russell & Russell, 1962.

   Translated from the Bridges edition. See item 186.

188. Baur, Ludwig. *Die philosophischen Werke des Robert Grosseteste, Bischofs von Lincoln*. *Beiträge zur Geschichte der Philosophie des Mittelalters*, 9. Münster: Aschendorff, 1912. Pp. xiii + 181 + 778.

   Editions of all the independent treatises, authentic or not.

189. Clagett, Marshall. "Some General Aspects of Physics in the Middle Ages." *Isis*, 39 (1948), 29-44.

   Traces the place of physical thought in the classification of the sciences, reviews medieval developments in optics, statics and magnetism and discusses ideas of motion in the fourteenth century, particularly the impetus theory.

190. Clagett, Marshall. *The Science of Mechanics in the Middle Ages*. Madison: University of Wisconsin Press, 1959. Pp. xxix + 711.

Traces the antecedents and scope of the central themes developed by medieval mechanicians, copiously illustrated with selections from the documents. While working within an essentially Aristotelian framework, the schoolmen criticized, modified and elaborated this legacy, and in so doing, formulated concepts in which lay the eventual refutation of Aristotle.

191. Clagett, Marshall. Studies in Medieval Physics and Mathematics. London: Variorum Reprints, 1979.

A collection of previously published articles. Includes items 109, 111, 189, 287, 288, 290, 291, 298, 839, 840 and 842.

192. Dales, Richard C. The Scientific Achievement of the Middle Ages. Philadelphia: University of Pennsylvania Press, 1973. Pp. ix + 182.

Presents selections from medieval authors and modern scholars within the framework of an historical and explanatory narrative on topics in medieval natural philosophy and astronomy.

193. DeKosky, Robert K. Knowledge and Cosmos: Development and Decline of the Medieval Perspective. Washington, D.C.: University Press of America, 1979. Pp. viii + 325.

Surveys the development and fate of astronomy and cosmology, theories of matter and conceptions of motion from antiquity to the scientific revolution. These topics are discussed within a four-fold framework of "traditional patterns of interpretation:"-- teleological, mechanical, occultist and mathematico-descriptive.

194. Delambre, Jean Baptiste Joseph. Histoire de l'astronomie du moyen âge. Paris, 1819. Reprint. New York: Johnson Reprint, 1966. Pp. lxxxiv + 640.

A survey of Arabic and European technical astronomy to the sixteenth century. Still valuable.

195. Dijksterhuis, E.J. The Mechanization of the World Picture. Translated by C. Dikshoorn. Oxford, 1961. Paperback reprint. Oxford: Oxford University Press, 1969. Pp. vi + 537.

Translated from De Mechanisering van het Wereldbeeld, Amsterdam, 1950. A survey of the development of physical and astronomical thought from antiquity through the seventeenth century. Part II (pp. 99-219) surveys medieval developments from the encyclopedists through the fourteenth century.

196. Duhem, Pierre. Les origines de la statique. 2 vols. Paris: Hermann, 1905-06.

In Vol. I explores the history of medieval statics, the "science of weights" and its authors, the auctores de ponderibus, especially the treatises ascribed to Jordanus de Nemore whom Duhem saw as a precursor of seventeenth-century mechanics. This pioneering study has been emended and reinterpreted on several points by Moody and Clagett. See item 255.

197. Duhem, Pierre. Études sur Léonard de Vinci. 3 vols. Paris, 1906, 1909, 1913. Reprint. Paris: de Nobele, 1955.

A pioneering work which deals for the most part with medieval physical thought as related to da Vinci's predecessors and his followers. Vol. I seeks the sources of Leonardo's ideas in Albert of Saxony, Themon Judaeus and the auctores de ponderibus. Vol. II discusses several themes explored at Paris and Oxford which provided a background for Leonardo. Vol. III treats reinterpretations of Aristotelian dynamics in the work of the fourteenth-century scholastics.

198. Duhem, Pierre. Le système du monde: Histoire des doctrines cosmologiques de Platon à Copernic. 10 vols. Paris: Hermann, 1913-1959.

A pioneering comprehensive work on cosmology, astronomy and physical thought from the Hellenistic period through the fifteenth century. Despite the corrections and reinterpretations inevitably due to subsequent research, despite Duhem's emphasis on French contributions and his tendency to search for medieval precursors of seventeenth-century giants, the work remains indispensable.

199. Franklin, Allan. The Principle of Inertia in the Middle Ages. Boulder: Colorado Associated University Press, 1976. Pp. xiv + 119.

Surveys views on motion from Aristotle through the seventeenth century.

200. Grant, Edward. Physical Science in the Middle Ages. New York: Wiley, 1971. Pp. xi + 128.

Surveys major topics in the history of physical thought including the condition of European science before the period of translation, the return of Greek materials to the West, the role of the university in the absorption of Aristotle, as well as medieval views on motion, astronomy and cosmology. Concludes with a long bibliographical essay.

201. Grant, Edward. Much Ado About Nothing: Theories of Space and Vacuum from the Middle Ages to the Scientific Revolution. Cambridge: Cambridge University Press, 1981. Pp. xiii + 456.

Traces the concept of void space from the thirteenth century through Newton. Scholastics were the heirs of Aristotle's denial of any kind of vacuum. While they supported his rejection of interstitial vacua, Aristotle's denial of the possibility of motion through empty space, led to speculations as to a hypothetical vacuum within the cosmos. In the fourteenth century, scholastics introduced an imaginary void space filled with the non-dimensional immensity of God. In Newton, void space, three-dimensional and infinite, was identified with the deity.

202. Grant, Edward. Studies in Medieval Science and Natural Philosophy. London: Variorum Reprints, 1981.

A collection of previously published articles. Includes items 13, pp. 265-302; 121, 299, 301, 302, 304, 305, 306, 456, 457, 662, 663, 664, 665 and 879.

203. Lindberg, David C. Theories of Vision from al-Kindī to Kepler. Chicago: University of Chicago Press, 1976. Pp. xii + 324.

Traces theories of the visual process from their origins in ancient and Hellenistic antiquity through the reception of this tradition in Islam, its transfer to the West and the high point of interest with the work of the perspectivists by the end of the thirteenth century. After a lean period in the late

Middle Ages, visual theories and perspective were revived in the sixteenth century, although dependent on a medieval tradition which reached its culmination with Kepler.

204. Maier, Anneliese. An der Grenze von Scholastik und Naturwissenschaft. 2d ed. Rome: Edizioni di storia e letteratura, 1952. Pp. vii + 388.

Treats the antecedents and medieval fortunes of three problems: The role of the elements and the primary qualities in the formation of mixtures, explanations of free fall and the causes of acceleration in free fall, the mathematical treatment of the latitude of forms doctrine in the work of the Mertonians and Nicole Oresme.

205. Maier, Anneliese. Metaphysische Hintergründe der spätscholastischen Naturphilosophie. Rome: Edizioni di storia e letteratura, 1955. Pp. viii + 405.

Presents essays which focus on the metaphysical component of the scholastic philosophy of nature, the concepts, principles and methods, rather than actual physical problems. Among the topics considered are the problem of time, discussions of quantity, forces and energies, final causes and nature's laws.

206. Maier, Anneliese. Zwischen Philosophie und Mechanik. Rome: Edizioni di storia e letteratura, 1958. Pp. vii + 385.

Essays primarily concerned with the late thirteenth and fourteenth century treatment of motion, a topic to which schoolmen contributed conceptions which re-interpreted or transcended their Aristotelian and Arabic sources.

207. Maier, Anneliese. Die Vorläufer Galileis im 14. Jahrhundert. 2d rev. ed. Rome: Edizioni di storia e letteratura, 1966. Pp. 367.

Argues that during the fourteenth century, scholastics constructed an independent natural philosophy which can be evaluated as the first stage of the eventual triumph over Aristotle and the development of classical physics.

In this volume, some of the issues involved are examined --fundamental concepts, mathematical and physical questions and ideological changes.

208. Maier, Anneliese. Zwei Grundprobleme der scholastischen Naturphilosophie. 3d rev. and exp. ed. Rome: Edizioni di storia e letteratura, 1968. Pp. 400.

Traces the sources and development of two major themes in scholastic natural philosophy: the intension and remission of forms and the theory of impetus as revealed in the thought of the chief natural philosophers of the late Middle Ages.

209. Maier, Anneliese. Ausgehendes Mittelalter: Gesammelte Aufsätze zur Geistesgeschichte des 14. Jahrhunderts. 3 vols. Rome: Edizioni di storia e letteratura, 1964, 1967, 1977.

A collection of previously published essays on medieval philosophy and natural philosophy (Vols. I & II). Vol. III, edited after Maier's death by Agostino Paravicini Bagliani, contains essays on the history of the papal library at Avignon and the controversy concerning the beatific vision during the pontificate of John XXII. Includes items 320, 321, 322, 323, 619, 620, 621, 622, 1067, 1244, 1434, 1435 and 1436.

210. Maier, Anneliese. On the Threshold of Exact Science: Selected Writings of Anneliese Maier on Late Medieval Natural Philosophy. Edited and translated with an introduction by Steven D. Sargent. Philadelphia: University of Pennsylvania Press, 1982. Pp. xiv + 173.

Has translations of selections from Die Vorläufer Galileis im 14. Jahrhundert, An der Grenze von Scholastik und Naturwissenschaft and Ausgehendes Mittelalter on motion, causes and forces, the concept of function, impetus and theories of matter. See items 204, 207 and 209.

211. Millás Vallicrosa, J.M. Assaig d'història de les idees físiques i matemàtiques a la catalunya medieval, I. Barcelona: Institució Patxot, 1931. Pp. xv + 349.

Surveys the scientific culture of medieval Catalonia from the introduction of classical materials in the Visigothic and Carolingian periods, through Arabic dominance, and during the Reconquest. The location of the Spanish March favored its important role in the early transmission of mathematical and astronomical information, especially of such instruments as the astrolabe and the quadrant, to the rest of Europe. The role of the monastery at Ripoll was pivotal in this transmission as was the work of the translator, Llobet of Barcelona. Several appendixes give texts of Llobet's astrolabe treatises and other material from Ripoll.

212. Millás Vallicrosa, J.M. Estudios sobre historia de la ciencia española. Barcelona: Consejo superior de investigaciones científicas, 1949. Pp. ix + 499.

Essays on topics in the history of Spanish science and technology to the Renaissance, chronologically ordered, several of which have been previously published. The crucial role played by medieval Spain in the transmission of Arabic science, especially astronomy, is illustrated by essays on the old quadrant, Arzachel and the Toledan Tables, Abraham ibn 'Ezra.

213. Millás Vallicrosa, J.M. Nuevos estudios sobre historia de la ciencia española. Barcelona: Consejo superior de investigaciones científicas, 1960. Pp. 365.

Represents a continuation of the author's 1949 Estudios with essays on diverse themes in Spanish science and technology, chronologically ordered. Although most have appeared previously in various journals, they are presented here in amplified form.

214. Pedersen, Olaf and Mogens Pihl. Early Physics and Astronomy: A Historical Introduction. New York: American Elsevier, 1974. Pp. 413.

Surveys developments in the physical and astronomical thought of the Greeks and Middle Ages; has a biographical appendix of major thinkers.

215. Schramm, Matthias. Ibn al-Haythams Weg zur Physik. Wiesbaden: Steiner, 1963. Pp. x + 348.

Argues that Alhazen successfully addressed the task of harmonizing Aristotle's doctrine with a mathematical treatment of astronomical and optical problems. While his combination of Aristotelian metaphysics and Ptolemaic kinematics did not involve experimental methods, Alhazen's use of experiment in his optical researches led to new conceptions concerning the nature of light.

216. Wallingford, Richard of. Richard of Wallingford: An Edition of his Writings. 3 vols. Edited with introductions, English translation and commentary by John D. North. Oxford: Clarendon Press, 1976.

Volume I contains the texts and translations of the mathematical, astrological and astronomical treatises as well as short ecclesiastical works written when Wallingford was abbot of St. Albans. Volume II provides a biographical sketch plus introductions and commentaries to the scientific works in volume I. The third volume contains illustrations, tables, appendixes and glossaries.

217. Weisheipl, James A. The Development of Physical Theory in the Middle Ages. Rev. ed. Ann Arbor: University of Michigan Press, 1971. Pp. 92.

Traces certain developments in physical theory from the early Middle Ages to the period of Galileo, with a particular focus on theories in the works of Albertus Magnus and Aquinas.

## PHYSICS (OPTICS EXCLUDED) AND NATURAL PHILOSOPHY BEFORE THE FOURTEENTH CENTURY

218. Aegidius Romanus. Egidii Romani Commentaria in octo libros Phisicorum. Venice, 1502. Reprint. Frankfort: Minerva, 1968.

219. Aegidius Romanus. Commentaria in libros de generatione et corruptione. Quaestiones super primo libro De generatione et corruptione. Venice, 1505. Reprint. Frankfort: Minerva, 1970.

220. Albertus Magnus. De caelo et mundo. Edited by Paul Hossfeld. Münster: Aschendorff, 1971. Pp. xxiv + 342.

221. Alexander of Aphrodisias. Commentaire sur les Météores d'Aristote par Alexandre d'Aphrodisias. Traduction de Guillaume Moerbeke. Edited by A.J. Smet. Louvain: Publications universitaires, 1968. Pp. cxxxiv + 526.

222. Aquinas, Thomas. In Aristotelis libros de caelo et mundo, de generatione et corruptione, meteorologicorum expositio. Edited by R.M. Spiazzi. Turin: Marietti, 1952. Pp. xxvii + 741.

223. Aquinas, Thomas. In octo libros De physico auditu sive Physicorum Aristotelis commentaria. Editio novissima. Edited by A.-M. Pirotta. Naples: M. d'Auria Pontificus Editor, 1953. Pp. 658.

224. Aquinas, Thomas. In octo libros physicorum Aristotelis expositio. Edited by P.M. Maggiòlo. Turin: Marietti, 1954. Pp. 663.

225. Aquinas, Thomas. "Saint Thomas Aquinas: On the Combining of the Elements." Translated by Vincent R. Larkin. Isis, 51 (1960), 67-72.

Translation of a letter addressed to a master Philip dealing with a problem arising from the De generatione as to how the forms of the elements remain in compounds. See item 204, p. 31f.

226. Aquinas, Thomas. Commentary on Aristotle's Physics. Translated by Richard J. Blackwell, Richard J. Spath and W. Edmund Thirlkel, with an introduction by Vernon J. Bourke. New Haven: Yale University Press, 1963. Pp. xxxii + 599.

227. Aquinas, Thomas. Expositio in libros Aristotelis De generatione et corruptione, continuatio per Thomam de Sutona. Edited by Francis E. Kelley. Munich: Verlag der Bayerischen Akademie der Wissenschaften, 1976. Pp. 211.

228. Aristotle. Physica. Translatio Vaticana. Edited by A Mansion. Aristoteles latinus, VII, 2. Bruges: Desclée de Brouwer, 1957. Pp. xvi + 43.

229. Averroes (Ibn Rushd). Commentarium medium in Aristotelis De generatione et corruptione libros. Edited by F.H. Fobes and S. Kurland. Cambridge, Mass.: Mediaeval Academy of America, 1956. Pp. xliv + 216.

230. Averroes (Ibn Rushd). On Aristotle's De generatione et corruptione. Middle Commentary and Epitome. Translated by Samuel Kurland. Cambridge, Mass.: Mediaeval Academy of America, 1958. Pp. xxv + 245.

Translated from the original Arabic and from the Hebrew and Latin versions.

231. Boetius of Dacia. Questiones de generatione et corruptione. Questiones super libros Physicorum. Edited by Géza Sajó. Copenhagen: Det Danske Sprogog Litteratureselskab (Gad), 1972-1974.

232. Boetius of Dacia. Questiones super IVm meteorologicorum. Edited by Gianfranco Fioravanti. Copenhagen: Gad, 1979. Pp. xxvi + 140.

233. Clagett, Marshall. "The Liber de motu of Gerard of Brussels and the Origins of Kinematics in the West." Osiris, 12 (1956), 73-175.

An edition of Gerard's treatise. Argues that it was the first on kinematics in western Europe. It had as its goal, the representation of non-uniform velocities by uniform ones as demonstrated in the cases of lines, surfaces and solids in rotation.

234. Dales, Richard C. "Robert Grosseteste's Scientific Works." Isis, 52 (1961), 381-402.

Considers a selection of Grosseteste's scientific works, arranged in their probable chronological order, with summaries of their content. Concludes that Grosseteste's scientific methodology developed in both his theoretical works and those on particular subjects.

235. Dales, Richard C. "The Authorship of the Questio de fluxu et refluxu maris Attributed to Robert Grosseteste." Speculum, 37 (1962), 582-588.

Argues that this is a genuine Grosseteste work based on MSS ascriptions and internal evidence. See items 240, 241.

236. Dales, Richard C. "Robert Grosseteste's Treatise *De finitate motus et temporis*." *Traditio*, 19 (1963), 245-266.

Points out in the introduction to the edition that this work, which refutes Aristotle's view of time and motion as eternal, exists in two versions. It was appended to Grosseteste's *Physics* commentary, but later separated; Grosseteste had intended it as an independent treatise.

237. Dales, Richard C. "The Authorship of the *Summa in physica* Attributed to Robert Grosseteste." *Isis*, 55 (1964), 70-74.

Argues from internal evidence and from comparison with the genuine *Commentarius in octo libros Physicorum* that the *Summa*, as Baur had thought and as Thomson had denied, is likely not by Grosseteste. See items 188, 263.

238. Dales, Richard C. "*Anonymi De elementis*: From a Twelfth-Century Collection of Scientific Works in British Museum: MS Cotton Galba E. IV." *Isis*, 56 (1965), 174-189.

Analyses the content of the treatise and gives the Latin text. Points out striking ideas such as an atomic theory of matter and perception, the eternity of the world and the earth's rotation. Discusses the sources used by the author, stressing his independence of thought, educational background and possible origin.

239. Dales, Richard C. "Richard de Staningtona: An Unknown Writer of the Thirteenth Century." *Journal of the History of Philosophy*, 4 (1966), 199-208.

Concludes from a study of de Staningtona's compilation of the *libri naturales*, written for arts instruction at an elementary level and his only known work, that the author was a Dominican who taught arts at the Oxford Dominican school around the mid-thirteenth century.

240. Dales, Richard C. "The Text of Grosseteste's _Questio de fluxu et refluxu maris_, with English Translation." _Isis_, 57 (1966), 455-474.

Argues that this work, written about 1227, shows familiarity with available theories on tides, but is also an original investigation of the subject. See items 235, 241.

241. Dales, Richard C. "Adam Marsh, Robert Grosseteste, and the Treatise on the Tides." _Speculum_, 52 (1977), 900-901.

Presents further research on the authorship of the treatise on tides tentatively ascribed to Grosseteste (_Speculum_, 1962). There is some evidence that the author may be Adam Marsh, friend and colleague of Grosseteste. See items 235, 240.

242. Dales, Richard C. "A Twelfth-Century Concept of the Natural Order." _Viator_, 9 (1978), 179-192.

Claims that the notion of nature as a self-sufficient, largely mechanical entity, was consciously held by several twelfth-century authors. One can see this development in Adelard of Bath at the beginning of the century to the precise expression of the idea in Urso of Calabria and John Blund.

243. Duhem, Pierre. "Roger Bacon et l'horreur du vide." _Roger Bacon Essays_. Edited by A.G. Little. Oxford: Clarendon Press, 1914, pp. 242-284.

Argues that Bacon based his rejection of a vacuum on the idea that the universal and fundamental nature of bodies assured their contiguity, a requirement prior to the laws governing their motions.

244. Dunphy, William. "The _quinque viae_ and Some Parisian Professors of Philosophy." _St. Thomas Aquinas, 1274-1974: Commemorative Studies_, II. Toronto: Pontifical Institute of Mediaeval Studies, 1974, pp. 73-104.

Discusses the use made by two Parisian scholastics during the years 1270-77 in their Aristotelian commentaries of Aquinas' proofs of the existence of God.

245. Effler, Roy R. John Duns Scotus and the Principle "Omne quod movetur ab alio movetur." St. Bonaventure, N.Y.: Franciscan Institute, 1962. Pp. xvi + 208.

Argues that Scotus rejected the universal nature of this principle and its implied denial of self-motion. In fact, Scotus applied the idea of self-motion to the natural motion of heavy and light bodies, to the local motion of animals, to qualitative and quantitative change and to the activity of the intellect and the will.

246. Grosseteste, Robert. Commentarius in VIII libros Physicorum. Edited by Richard C. Dales. Boulder, Colo.: University of Colorado Press, 1963. Pp. xxxii + 192.

247. Hossfeld, Paul. "Die naturwissenschaft/naturphilosophische Himmelslehre Alberts des Grossen (nach seinem Werk De caelo et mundo)." Philosophia naturalis, 11 (1969), 318-359.

Describes the major themes of Albert's commentary on the De caelo. The work is a paraphrase of the Aristotelian text with digressions which relied on translations from Arabic by Gerard of Cremona and Michael Scot. See item 220.

248. Hossfeld, Paul. "Die Ursachen der Eigentümlichkeiten der Elemente nach Albertus Magnus." Philosophia naturalis, 14 (1973), 197-209.

Discusses the major themes in Albert's widely-read commentary on the pseudo-Aristotelian De causis proprietatum elementorum.

249. Hossfeld, Paul. "Grundgedanken in Alberts des Grossen Schrift Über Entstehung und Vergehen." Philosophia naturalis, 16 (1977), 191-204.

Summarizes the content of Albert's De generatione commentary, a work which follows the Aristotelian text closely.

250. Jansen, B. "Olivi der älteste scholastische Vertreter der heutigen Bewegungsbegriffs." Philosophisches Jahrbuch der Görresgesellschaft, 33 (1920), 137-152.

Argues that Peter John Olivi's explanation of projectile motion as an impression or inclination introduced into the moving body which moves it further in the absence of an accompanying mover was a precursor of the fourteenth-century theory of impetus.

251. Kurdziałek, Marian. "David von Dinant und die Anfänge der aristotelischen Naturphilosophie." La filosofia della natura nel medioevo. Atti del terzo congresso internazionale di filosofia mediovale, 1964. Milan: Vita e pensiero, 1966, pp. 407-416.

Argues that David of Dinant was familiar with Aristotle's works before Albertus Magnus and Thomas Aquinas, and was the first in the West to use this material to create his own philosophical system.

252. McCullough, Ernest J. "St. Albert on Motion as Forma fluens and Fluxus formae." Albertus Magnus and the Sciences: Commemorative Essays. Edited by James A. Weisheipl. Toronto: Pontifical Institute of Mediaeval Studies, 1980, pp. 129-153.

Argues from a detailed consideration of Albert's treatment of Physics III, 1-3 that Maier is mistaken in claiming Albert as the source for the distinction between these two interpretations of the nature of motion. Moreover, Maier has misunderstood Albert's discussion of the role of Avicenna as the source of the fluxus formae notion, and has over-simplified Averroes' position in regard to motion as a forma fluens. See items 206, p. 73f and 207.

253. McKeon, Charles K. A Study of the Summa philosophiae of the Pseudo-Grosseteste. New York: Columbia University Press, 1948. Pp. xi + 226.

Surveys the content of this thirteenth-century philosophical encyclopedia which reveals a strong relationship to the thought of Grosseteste and Roger Bacon. Argues that the treatise was probably written by an Englishman, possibly Kilwardby, between 1265-1275. The final six tracts of the Summa deal with natural philosophy in which the author treats light and nature as principles of material entities.

254. Moody, Ernest A. "Galileo and Avempace: The Dynamics of the Leaning Tower Experiment." Studies in Medieval Philosophy, Science, and Logic: Collected Papers, 1933-1969. Berkeley: University of California Press, 1975, pp. 203-286.

Maintains that the views on dynamics held by Galileo in his Pisan period (the time of the supposed Leaning Tower experiment) rejected both Aristotle's dynamics and the fourteenth-century interpretations of Aristotle of Bradwardine and Buridan, and are likewise far different from Galileo's mature opinion. His Pisan dynamics was based on Avempace as reported by Averroes, and on the sixth-century Neoplatonic Aristotelian critic, Philoponus.

255. Moody, Ernest A. and Marshall Clagett. The Medieval Science of Weights. Madison: University of Wisconsin Press, 1960. Pp. x + 438.

Presents editions with English translations of eight treatises on statics ascribed to Euclid, Archimedes, Thābit ibn Qurra, Jordanus de Nemore and Blasius of Parma. The medieval science of statics, the science of weights (scientia de ponderibus) was among the subgroups of physics which stemmed from the Hellenistic mathematical tradition, amplified and transmitted by Arabic intermediaries. Fed by Greek and Arabic sources, statics reached its fruition in the thirteenth century with the work of Jordanus de Nemore.

256. Peter of Maricourt. The Letter of Petrus Peregrinus On the Magnet, A.D. 1269. Translated by Brother Arnold (Joseph Charles Mertens). Introduction by Brother Potamian (M.F. O'Reilly). New York: McGraw-Hill, 1904. Pp. iii-xix + 41.

257. Peter of Maricourt. "Le De magnete de Pierre de Maricourt." Translated with a commentary by P. Radelet-de Grave and D. Speiser. Revue d'histoire des sciences, 28 (1975), 193-234.

Text and French translation with a commentary in the footnotes.

258. Schlund, Erhard. "Petrus Peregrinus von Maricourt: Sein Leben und seine Schriften." Archivum

Franciscanum Historicum, 4 (1911), 436-455; 5 (1912), 22-40.

Provides information on Peter's life (Part I); lists MSS of the Epistola de magnete (Part II) and printed editions (Part III).

259. Siger of Brabant. Questions sur Physique d'Aristote. Edited by Philippe Delhaye. Louvain: Institut supérieur de philosophie, 1941. Pp. 254.

See items 1413, 1434 and 1435.

260. Siraisi, Nancy G. "The Expositio problematum Aristotelis of Peter of Abano." Isis, 61 (1970), 321-339.

Analyzes the structure of the commentary of Peter of Abano (ca. 1250-ca. 1316) on the pseudo-Aristotelian Problemata which he based on the translation made by Bartholomew of Messina. Points out Peter's emphasis on secondary causation, his readiness to question authority and his skepticism. Peter's treatment of astrology, medicine, the quadrivium, natural science and ethics are treated in detail.

261. Steele, Robert. "Practical Chemistry in the Twelfth Century." Isis, 12 (1929), 10-46.

Gives the text of the De aluminibus et salibus of Rhazes (al-Rāzi) which was translated by Gerard of Cremona. The work deals with the preparation of chemical substances divided into four classes: alums or vitriols, salts, volatile agents and "bodies."

262. Théry, G. Autour du decret de 1210, II: Alexandre l'Aphrodise aperçu sur l'influence de sa noétique. Paris: Le Saulchoir, Kain, 1926. Pp. 120.

Editions of Alexander's De tempore, De sensu and Quod augmentum et incrementum fuit in forma et non in yle. These were translated by Gerard of Cremona.

263. Thomson, S. Harrison. "The Summa in VIII libros Physicorum of Grosseteste." Isis, 22 (1934), 12-18.

Argues, by an analysis of the purpose and content of the treatise as well as new MS evidence, that previous opinion as to the spurious attribution of the Summa is unjustified. See item 237.

264. Thomson, S. Harrison. "Grosseteste's Questio de calore, De cometis and De operacionibus solis." Medievalia et Humanistica, 11 (1957), 34-43.

Provides texts of the Questio de calore, a university exercise on the term calor, likely written ca. 1230, a more complete version of the De cometis than that published in 1933, and a De operacionibus solis (not titled in the MS). See items 188, 500.

265. Thorndike, Lynn. "John of St. Amand on the Magnet." Isis, 36 (1939), 156-157.

Points out that although Gilbert made no mention of him, John (fl. 1261-1298) wrote on the magnet in his commentary on the Antidotarium Nicolai. Contains the text and translation of the passage.

266. Thorndike, Lynn. Michael Scot. London: Thomas Nelson, 1965. Pp. 143.

Discusses the life, career and principal features of Scot's thought as revealed primarily in his Liber introductorius. Treats likewise Scot's role as a translator of Aristotle and Arabic authors.

267. Wallace, William A. "Gravitational Motion according to Theodoric of Freiberg." The Dignity of Science. Edited by James A. Weisheipl. Washington, D.C.: Thomist Press, 1961, pp. 191-216.

Discusses the views on the motion of bodies expressed in Theodoric's De elementis corporum naturalium inquantum sunt partes mundi.

268. Weisheipl, James A. "Natural and Compulsory Movement." The New Scholasticism, 29 (1955), 50-61.

Points out that based on Aristotle scholastics distinguished between a natural motion which implied an "inner intention" characteristic of the body, and a compulsory or violent motion which involved no such intention, but was forced on the body from an external source. The impetus theory, advanced to account for the violent motion of projectiles, and which even when permanent was overcome by the resistance of the medium and the natural tendency of the body, was developed within an Aristotelian context.

269. Weisheipl, James A. "The Principle Omne quod novetur ab alio movetur in Medieval Physics." Isis, 56 (1965), 26-45.

Argues that this principle, fundamental to Aristotelian and medieval explanations of natural motion, that of projectiles and of celestial motion has been often misunderstood by both scholastics and modern historians. It was not Aristotle's intent that the form of a body undergoing motion be interpreted as an accompanying mover (a motor coniunctus). The principle was thus misunderstood by Averroes, but Aquinas interpreted Aristotle correctly.

270. Weisheipl, James A. "The Commentary of St. Thomas on the De caelo of Aristotle." Sapientia, 29 (1974), 11-34.

Points out that this work, one of the last of Thomas' writings, deals in a mature way with the topics adduced by Aristotle and his Greek and Arabic commentators, especially Simplicius and Averroes. Thomas used the De caelo translation from the Greek as well as the translation of Simplicius' commentary made by his friend Moerbeke, a version which became the standard text. Although Thomas was aware of the existence of the entire text, his commentary stops within Book III; likely he was prevented from completing the commentary by his final illness. Summarizes the most important features of the De caelo, and particularly Thomas' interpretations of the nature of celestial movers and the natural motion of terrestrial bodies.

271. Weisheipl, James A. "Motion in a Void: Aquinas and Averroes." St. Thomas Aquinas, 1274-1974: Commemorative Studies, I. Toronto: Pontifical Institute of Mediaeval Studies, 1974, pp. 468-488.

Maintains that while Averroes interpreted one of Aristotle's arguments against the atomists in the Physics as implying that the Greek master himself believed that a resisting medium was necessary for natural motion, Philoponus, Avempace and St. Thomas are representatives of a tradition which did not agree.

272. Weisheipl, James A. "The Spector of Motor coniunctus in Medieval Physics." Studi sul XIV secolo in memoria Anneliese Maier. Edited by Alfonso Maierù and Agostino Paravicini Bagliani. Rome: Edizioni di storia e letteratura, 1981, pp. 81-104.

Argues that the Aristotelian dictum: Omne quod movetur ab alio movetur has been misinterpreted to mean that an efficient cause, a motor coniunctus, is required for natural motion. However, Aristotle intended that the substantial form, as a principle, was sufficient to account for natural motion and he was so interpreted by Albertus Magnus and Aquinas, although Averroes and Avicenna understood Aristotle wrongly. Duns Scotus, to whom all motion was self-moving, rejected the dictum.

273. Weisheipl, James A. "Aristotle's Concept of Nature: Avicenna and Aquinas." Approaches to Nature in the Middle Ages. Edited by Lawrence D. Roberts. Binghamton, N.Y.: Center for Medieval and Early Renaissance Studies, 1982, pp. 137-157.

Argues that while both Avicenna and Aquinas considered themselves Aristotelians, Avicenna was, in fact, a Platonist. As a result there were fundamental differences between them as to the actual determinant of natural motion, in their explanations of substantial change, and in their attitudes toward determinism.

274. Zimmermann, Albert, ed. Ein Kommentar zur Physik des Aristoteles aus der Pariser Artistenfakultät um 1273. Berlin: de Gruyter, 1968. Pp. l + 106.

An edition of a fragment of a commentary on Book I with excerpts from Books IV and VIII from a MS in the Bibliothèque Nationale. The material reveals the influence of both Aquinas and Siger of Brabant; the author was surely among those arts masters writing between 1271-1277 who were affected by the events of 1277.

## NATURAL PHILOSOPHY OF THE FOURTEENTH CENTURY

275. Albert of Saxony. "Der *Tractatus proportionum* von Albert von Sachsen." Edited by H.L.L. Busard. *Österreichische Akademie der Wissenschaften, math.-natur. Klasse. Denkschriften*, 116, 2 (1971), 43-72.

An edition of Albert's work. Kinematic and dynamic treatments of motion were explored in the West by Gerard of Brussels and the Mertonians, particularly Bradwardine, and were taken up in Paris by Buridan, Oresme and Albert. The latter's treatise was probably composed between 1353 and 1365. It reveals a strong dependency on Bradwardine's *Tractatus de proportionibus* as well as on other Mertonians and Oresme.

276. Baudry, Léon. "Les rapports de Guillaume d'Occam et de Walter Burleigh." *Archives d'histoire doctrinale et littéraire du moyen âge*, 9 (1934), 155-173.

Argues that one cannot assume a direct confrontation between Burley and Ockham, presumably beginning at Oxford and continuing when Burley left for Paris. It is not certain that Ockham, in his attacks on the opinions of his opponents, had Burley explicitly in mind nor that he was refuting Burley directly.

277. Baudry, Léon. *Lexique philosophique de Guillaume d'Ockham*. Paris: P. Lethielleux, 1958. Pp. ii + 298.

Lists important philosophical terms and their various nuances of meaning drawn from Ockham's works. These are frequently illustrated by quotations from the texts.

278. Borchert, Ernst. *Die Lehre von der Bewegung bei Nicolaus Oresme*. *Beiträge zur Geschichte der Philosophie und Theologie des Mittelalters*, 31, 3. Münster: Aschendorff, 1934. Pp. xvi + 112.

Discusses Oresme's views on motion based on passages from the *Du ciel*, and on time (based on the *De configurationibus*). Argues that Oresme saw motion as a flux in relation to an imagined immobile space identified with the infinity of God, that he interpreted time as a modus of things, and that Oresme's view of impetus as non-permanent and related to acceleration represented an advance over the positions taken by his contemporaries.

279. Bradwardine, Thomas. Thomas of Bradwardine, Tractatus de proportionibus: Its Significance for the Development of Mathematical Physics. Edited and translated by H. Lamar Crosby, Jr. Madison: University of Wisconsin Press, 1955. Pp. viii + 203.

Written in 1328, Bradwardine's work, including his enormously influential exponential rule, reinterpreting Aristotle's dynamics to mean that as velocity increases arithmetically, the ratio of force to resistance increases geometrically, marked the beginning of the application of mathematics to natural philosophy.

280. Brampton, C. Kenneth. "Ockham and his Authorship of the Summulae in libros physicorum." Isis, 55 (1964), 418-426.

Argues on the basis of internal evidence and a comparison with other Ockham treatises that the Summulae is not a genuine Ockham work, despite the consensus of Ockham scholarship to the contrary.

281. Breidert, Wolfgang. Das aristotelische Kontinuum in der Scholastik. Beiträge zur Geschichte der Philosophie und Theologie des Mittelalters, n.f., 1. Münster: Aschendorff, 1970. Pp. viii + 76.

Traces the philosophical background of the history of the continuum from the paradoxes of Zeno to the seventeenth century, focusing on motion, the central feature of Aristotelian natural philosophy and related by him to the continuity of space and time. Within this context, surveys scholastic interpretations and modifications of Aristotle's doctrines on the continuum and the nature of the infinite.

282. Buridan, John. Questiones super octo Physicorum libros Aristotelis. Paris, 1509. Reprint. Frankfort: Minerva, 1964.

283. Buridan, John. Quaestiones super libris quattuor de caelo et mundo. Edited by Ernest A. Moody. Cambridge, Mass., 1942. Reprint. New York: Kraus, 1970. Pp. xxxv + 274.

284. Burley, Walter. In Physicam Aristotelis expositio et quaestiones. Venice, 1501. Reprint. Hildesheim: Olms, 1972.

285. Busard, H.L.L. "Die Quellen von Nicole Oresme." Janus, 58 (1971), 162-193.

Traces the numerous mathematical-physical sources cited by or referred to in Oresme's mathematical-physical treatises, especially the De configurationibus, and investigates the routes of transmission of the mathematical ideas used by Oresme in translations from Greek or Arabic, or which were transmitted by Oresme's more immediate predecessors.

286. Clagett, Marshall. Giovanni Marliani and Late Medieval Physics. New York: Columbia University Press, 1941. Pp. 182.

Considers three areas in natural philosophy and mathematics explored in Marliani's works: heat actions, problems in mechanics, and in mathematics, fractions. Concludes that Marliani was familiar with the work of his English and French predecessors, but although superior to his contemporaries, did not transcend their accomplishments.

287. Clagett, Marshall. "Richard Swineshead and Late Medieval Physics. I: The Intension and Remission of Qualities." Osiris, 9 (1950), 131-161.

Argues that the treatment of qualities in a quantitative fashion in which an increase took place by degree to degree addition introduced mathematical considerations into natural philosophy in the fourteenth century, as shown in the first tract of Swineshead's Calculations.

288. Clagett, Marshall. "The Use of Points in Medieval Natural Philosophy and most particularly in the Questiones de spera of Nicole Oresme." Studies in Medieval Physics and Mathematics. London: Variorum Reprints, 1979, pp. 215-221.

Argues that in Oresme's thought certain points, such as the center of gravity of the earth and the center of the universe, could be considered fictional entities within the context of medieval discussions of points in general. But the use of these two points, in the context of the possible, continual oscillation of the earth through the center of the universe, gave these concepts a kind of physical reality. Reprinted from Actes du Symposium International R.J. Bošković, 1961. Belgrade, 1962.

289. Clagett, Marshall. "Nicole Oresme and Medieval Scientific Thought." Proceedings of the American Philosophical Society, 108 (1964), 298-309.

Reviews the life, career and work of this highly original schoolman, with emphasis on his configuration doctrine. See item 353.

290. Clagett, Marshall. "The Pre-Galilean Configuration Doctrine: 'The Good Treatise on Uniform and Difform [Surfaces]'." Studies in Medieval Physics and Mathematics. London: Variorum Reprints, 1979, pp. 1-24.

Presents, within the context of Galileo's use of the configuration doctrine, the text of a treatise on uniform and uniformly difform surfaces in an Erfurt MS, dependent on Casali and/or Oresme's Questions on Euclid. Reprinted from Pubblicazioni de Comitato nazionale per le manifestazioni celebrativi, III. Saggi su Galileo Galilei. Edited by Carlo Maccagni, I. Florence, 1967.

291. Clagett, Marshall. "Some Novel Trends in the Science of the Fourteenth Century." Art, Science, and History in the Renaissance. Edited by C.S. Singleton. Baltimore: Johns Hopkins Press, 1967, pp. 275-303.

Traces some of the advances in fourteenth-century thought that went beyond the heritage of Aristotelian natural philosophy particularly the developments at Merton and the subsequent work of Oresme.

292. Clagett, Marshall. "Prosdocimus de Beldomandis and Nicole Oresme's Proof of the Merton Rule of Uniformly Difform." Isis, 60 (1969), 223-225.

Provides the text of an exposition of the proof of the Merton rule by this early fifteenth-century natural philosopher. This is the only extant geometrical writing of this author.

293. Corvino, Francesco. "Le _Quaestiones in libros physicorum_ nella formazione del pensiero di Guglielmo d'Occam." _Rivista critica di storia della filosofia_, 12 (1957), 385-411.

Argues that a comparison of Ockham's treatment of several themes and problems found in his earlier works with that in his _Questiones in libros physicorum_ (one of Ockham's last treatises) reveals a trend toward the simplification of arguments, increasing rigor and greater coherency.

294. Drake, Stillman. "Medieval Ratio Theory versus Compound Medicines in the Origins of Bradwardine's Rule." _Isis_, 64 (1973), 67-77.

Argues that while the source of Bradwardine's exponential rule may have been the quantitative pharmacy of Arnald of Villanova and the Montpellier school, his use of this notion was not a mere extension of its application to dynamics, but was a creative employment of the medieval theory of proportions not implied in the medical treatment. See item 948.

295. Eldredge, Laurence. "Late Medieval Discussions of the Continuum and the Point of the Middle English _Patience_." _Vivarium_, 17 (1979), 90-115.

Discusses fourteenth-century controversies between representative atomists and their opponents who defended the infinite divisibility of _continua_ within the context of the meaning of the word "point" in this poem.

296. Ermatinger, Charles J. "Urbanus Averroista and Some Early Fourteenth-Century Philosophers." _Manuscripta_, 11 (1967), 3-38.

Points out that Urbanus Bononiensis Averroista in his _Expositio_ of Averroes' commentary on the _Physics_, likely written at Bologna in 1334, makes considerable use of authors connected with early fourteenth-century Bologna such as Gentile da Cingoli, William of Alnwick, John of Jandun and Walter Burley.

297. Federici Vescovini, Graziella. Astrologia e scienza: La crisi dell'aristotelismo sul cadere del Trecento e Biagio Pelacani da Parma. Florence: Vallecchi, 1979. Pp. 472.

Surveys Blasius' work in natural philosophy, optics, logic and mathematics. Concludes that in his criticism of much of the Aristotelianism of his day and in his reduction of religious truth to natural astrology, Blasius was a daring and original thinker, many of whose ideas were not acceptable to his contemporaries.

298. Francesco of Ferrara. "Francesco of Ferrara's Questio de proportionibus motuum." Edited by Marshall Clagett. Annali dell'Istituto e Museo di storia della scienze di Firenze, 3 (1978), 3-63.

An edition of the Questio. It illustrates the spread of English treatises on the quantitative treatment of qualities and motions, especially Bradwardine's Tractatus de proportionibus. Francesco taught at Padua and determined this Questio there in 1352.

299. Grant, Edward. "Jean Buridan: A Fourteenth-Century Cartesian." Archives internationales d'histoire des sciences, 10 (1963), 251-255.

Points out that Buridan like Descartes identified the matter of a body and its dimensions with space. While both Buridan and Descartes distinguished place from space, they differed in their interpretations of the meaning of place.

300. Grant, Edward. "Aristotle's Restriction on his Law of Motion: Its Fate in the Middle Ages." Mélanges Alexandre Koyré, I. Paris: Hermann, 1964, pp. 173-197.

Argues that Aristotle's limitation to his rules of motion (exemplified in the shiphauler argument) was understood by scholastics before Bradwardine's exponential rule. After Bradwardine, those who accepted this new function either ignored Aristotle's restriction or interpreted it in odd ways.

301. Grant, Edward. "Motion in the Void and the Principle of Inertia in the Middle Ages." Isis, 55 (1964), 265-292.

Argues that one of the consequences Aristotle adduced against a void--an inertial motion--was not discussed by scholastics, although other implications of Aristotle's rejection of motion in a vacuum did lead to anti-Aristotelian views. Suggests that scholastics tended to focus on these other consequences, and that in a finite world, infinite motion would be thought impossible.

302. Grant, Edward. "Bradwardine and Galileo: Equality of Velocities in the Void." Archive for History of Exact Sciences, 2 (1965), 344-364.

Argues that before Benedetti and Galileo Bradwardine had maintained that bodies of the same material of different weights and sizes would fall with equal speeds in a void. Scholastics had assumed the anti-Aristotelian position that motion in a hypothetical void could be successive. In the case of mixed bodies (composed of two or more of the elements) this motion could be finite assuming a semi-quantitative internal resistance due to the ratio of degrees of heaviness and lightness of the elements in the body. Bradwardine concluded that unequal but homogenous bodies would fall in a void with equal speeds.

303. Grant, Edward. "The Arguments of Nicholas of Autrecourt for the Existence of Interparticulate Vacua." Actes du XIIe congrès international d'histoire des sciences, IIIa. Paris: Blanchard, 1971, pp. 65-68.

Argues that Nicholas of Autrecourt may have been the only scholastic to believe in the existence of indivisible atoms and interparticulate vacua. He held this view as part of a physical theory that saw all existing things as eternal.

304. Grant, Edward. "Medieval Explanations and Inerpretations of the Dictum that 'Nature Abhors a Vacuum'." Traditio, 29 (1973), 327-355.

Points out that all scholastics, with the possible exception of Nicholas of Autrecourt, denied the existence of any sort of separate, continuous, extended vacuum. Two "experiments" usually adduced to illustrate this impossibility--the clepsydra and the separation of two surfaces--led to ingenious arguments by scholastics which ranged from a reliance on a universal nature of celestial origin which overrode the natural motion of bodies to Blasius of Parma's recourse to logical doctrines.

305. Grant, Edward. "The Concept of ubi in Medieval and Renaissance Discussions of Place." Science, Medicine and the University: 1200-1550. Essays in Honor of Pearl Kibre, I. Edited by Nancy G. Siraisi and Luke Demaitre. Manuscripta, 20 (1976), pp. 71-80.

Summarizes the positions taken by scholastics concerning the meaning of place as applied to physical bodies from the thirteenth century to the early modern period as discussed in the lengthy treatment of the subject by Francisco Suarez (1548-1617). There were six major views, five of which can be divided into two groups: those who argued that ubi was external to the body and those who viewed it as internal. A completely different interpretation was held by Ockham who denied the reality of ubi as distinct from a body and the place of a body.

306. Grant, Edward. "The Principle of the Impenetrability of Bodies in the History of Concepts of Separate Space from the Middle Ages to the Seventeenth Century." Isis, 69 (1978), 551-571.

Traces the thought of scholastics who rejected the Aristotelian denial of a separate three-dimensional void space identified with body based on the impossibility of the interpenetration of bodies. The issue was involved with the problem of whether the essential determinant of the motion of bodies was resistance or the time taken to traverse space.

307. Grant, Edward. "The Medieval Doctrine of Place: Some Fundamental Problems and Solutions." Studi sul XIV secolo in memoria Anneliese Maier. Edited by Alfonso Maierù and Agostino Paravicini Bagliani. Rome: Edizioni di storia e letteratura, 1981, pp. 57-79.

Presents responses of scholastics to the troublesome Aristotelian heritage concerning place, particularly the distinction between proper place and common place, the immobility of place, and the place (possibly) of the ultimate sphere of the world.

308. Hall, Bert S. "The Scholastic Pendulum." Annals of Science, 35 (1978), 441-462.

Argues that despite the great antiquity of pendular motion, treatment of the pendulum in a mechanical context did not appear in the West until the fourteenth century work of Buridan and Oresme. Differing from explanations of this "discovery" offered by Thomas Kuhn and Piero Ariotti, suggests that this was due to the extension of Buridan's impetus theory to many instances of the persistence of motion drawn from everyday experience, particularly the motion of the heavy suspended bell characteristic of European Christianity. A further step was taken in Oresme's exploitation of the oscillatory motion of a heavy body dropped through an imaginary hole perforating the center of the earth. Use of this image, introduced previously by Adelard of Bath, in the context of Aristotelian cosmological and physical theories, became with Oresme an abstract treatment of the pendulum.

309. Hoskin, M.A. and A.G. Molland. "Swineshead on Falling Bodies: An Example of Fourteenth-Century Physics." British Journal for the History of Science, 3 (1966), 150-182.

Provides an example of the mathematical skill of this Mertonian revealed in his Liber calculationum. The problem arises in the context of whether or not the whole of a body is equal to the sum of its parts in the case of a body falling through an imaginary hole through the center of the earth when the body has fallen past the center and is moving upward. Following the proportionality laws established by Bradwardine, Swineshead compares the proportion of force to resistance in terms of the ratios of the lengths of the body on either side of the center. He arrives at the paradoxical result that in no finite time will the center of the body coincide with the center of the earth.

310. John of Jandun. Joannes de Janduno Super octo libros Aristotelis de physico auditu subtillissimae quaestiones. Venice, 1551. Reprint. Frankfort: Minerva, 1969.

311. Junghans, Helmar. Ockham im Lichte der neueren Forschung. Berlin: Lutherisches Verlaghaus, 1968. Pp. 367.

Discusses the results of recent scholarship, particularly that of Boehner, on questions concerning Ockham's life and career, the authenticity and chronology of his works and his philosophical position.

312. Knuuttila, Simo and Anja Inkeri Lehtinen. "Change and Contradiction: A Fourteenth Century Controversy." Synthèse, 40 (1979), 189-207.

Given in Aristotle that change from one state to another occurs instantaneously, and that time is continuous, it would appear that change involved a break in the law of contradiction. Discusses the thought of fourteenth-century adherents of this view (that instantaneous change implied contradiction), chiefly Landulf Caraccioli. These scholastics accounted for the dilemma by explaining that the temporal instant containing the contradictories could be logically divided into instants of nature. See item 315.

313. Koyré, Alexandre. "Le vide et l'espace infini au XIVe siècle." Archives d'histoire doctrinale et littéraire du moyen âge, 17 (1949), 45-91.

Argues that the possibilities offered by the two "cosmological" condemnations in the 1277 list at Paris--God can move the entire universe in rectilinear motion, a vacuum remaining, and God can create more than one world--both implying a void outside the cosmos, were not exploited by scholastics, given their absurdity in an Aristotelian cosmology. But Bradwardine specifically employed the notion of an infinite extracosmic void filled with an omnipresent God.

314. Kretzmann, Norman. "Richard Kilvington and the Logic of Instantaneous Speed." Studi sul XIV secolo in memoria Anneliese Maier. Edited by Alfonso Maierù and Agostino Paravicini Bagliani. Rome: Edizioni di storia e letteratura, 1981, pp. 143-178.

Provides the text and translation of sophisms 33 and 34 from Kilvington's Sophismata (written after Heytesbury's Regule) which contain an interpretation of instantaneous velocity quite different from that of Heytesbury and closer to the definition in modern kinematics.

315. Kretzmann, Norman. "Continuity, Contrariety, Contradiction, and Change." Infinity and Continuity in Ancient and Medieval Thought. Edited by Norman Kretzmann. Ithaca: Cornell University Press, 1982, pp. 270-296.

Argues that the theory of change which accepted contradiction at the instant of translation (apparently arising from Aristotle) adopted by some fourteenth-century scholastics, called here "quasi-Aristotelians," and which relied on the notion of the divided instant, was a response to a pseudo-problem based on a misinterpretation of Aristotle. Richard Kilvington treated the dilemma of simultaneous contradictory states in paradoxes which he resolved by Aristotelian analysis. See item 312.

316. Kretzmann, Norman, ed. Infinity and Continuity in Ancient and Medieval Thought. Ithaca: Cornell University Press, 1982. Pp. 367.

In addition to the essays, this volume contains appendixes with edited material from texts by Walter Burley and Richard Kilvington. Includes items 315, 340, 360 and 365.

317. Lappe, Joseph. Nicolaus von Autrecourt, sein Leben, sein Philosophie, seine Schriften. Beiträge zur Geschichte der Philosophie des Mittelalters, 6. Münster: Aschendorff, 1908. Pp. 48.

An older work analyzing Autrecourt's thought. Has the text of two of his letters to Bernard of Arezzo and part of the correspondence between Nicholas and a certain Egidius.

318. Leff, Gordon. Gregory of Rimini: Tradition and Innovation in Fourteenth-Century Thought. Manchester: Manchester University Press, 1961. Pp. x + 245.

An extensive survey of Gregory's views on epistemology, theology and natural philosophy based on his commentary on the Sentences. Argues that Gregory's thought should be considered in the context of the fourteenth century so as to determine to what extent his thought was nominalist or Augustinian, traditional or "modern."

319. Leff, Gordon. William of Ockham: The Metamorphosis of Scholastic Discourse. Manchester: Manchester University Press, 1975. Pp. xxiv + 666.

A survey embracing all aspects of Ockham's thought.

320. Maier, Anneliese. "Diskussionen über das actuell Unendliche in der ersten Hälfte des 14. Jahrhunderts." Ausgehendes Mittelalter: Gesammelte Aufsätze zur Geistesgeschichte des 14. Jahrhunderts, I. Rome: Edizioni di storia e letteratura, 1964, pp. 41-83.

Points out that while most fourteenth-century scholastics held to tradition and rejected the possibility of an actual infinite, there were several proponents of this position, some of whose speculations regarding the nature of the actual infinite are surprisingly modern. Reprinted from Divus Thomas, ser. 3, 24 (1947), 147-166, 317-337.

321. Maier, Anneliese. "Die naturphilosophische Bedeutung der scholastischen Impetustheorie." Ausgehendes Mittelalter: Gesammelte Aufsätze zur Geistesgeschichte des 14. Jahrhunderts, I. Rome: Edizioni di storia e letteratura, 1964, pp. 353-379.

Raises the question as to whether Buridan's version of a permanent impetus applied to the circular motion of the heavens can be considered as a foreshadowing of the principle of inertia. Concludes that while scholastic ideas concerning the nature of motion provide no terrestrial analogue to classical mechanics, Buridan's special case can be seen as an anticipation of the inertial principle. Reprinted from Scholastik, 30 (1955), 321-343.

322. Maier, Anneliese. "Die Stellung der scholastischen Naturphilosophie in der Geschichte der Physik."

Ausgehendes Mittelalter: Gesammelte Aufsätze zur Geistesgeschichte des 14. Jahrhunderts, I. Rome: Edizioni di storia e letteratura, 1964, pp. 413-424.

Argues that late scholastic natural philosophy certainly prepared the way for classical physics, although it did not anticipate it. Fourteenth-century scholastics at Paris and Oxford created a new natural philosophy, but more important and astonishing was the philosophical and methodological background to this new physics. Reprinted from Aus der deutschen Forschung der letzten Dezennien (Festschrift Ernst Telschow), Stuttgart, 1956, pp. 33-40.

323. Maier, Anneliese. "'Ergebnisse' der spätscholastischen Naturphilosophie." Ausgehendes Mittelalter: Gesammelte Aufsätze zur Geistesgeschichte des 14. Jahrhunderts, I. Rome: Edizioni di storia e letteratura, 1964, pp. 425-457.

Argues that it is ahistorical to inquire if the physical theories of late scholasticism were "correct," nor is it possible to determine to what extent the fourteenth century arrived at "results" that anticipated classical physics. The real achievement of late scholastic natural philosophy lay in new formulations of scientific methodology which did not, however, result in the development of quantitative science. Reprinted from Scholastik, 35 (1960), 161-188.

324. Markowski, Mieczysław. "Die philosophischen Richtungen in der Naturphilosophie an der Krakauer Universität in der ersten Hälfte des XV. Jahrhunderts." La filosofia della natura nel medioevo. Atti del terzo congresso internazionale di filosofia medioevale, 1964. Milan: Vita e pensiero, 1966, pp. 662-669.

Discusses some of commentaries on the libri naturales, De anima and the Parva naturalia written by Cracow arts masters and in the Jagellonian Library at the University. The predominating influence on these commentators was Buridan and his "school," Nicole Oresme, Albert of Saxony and Marsilius of Inghen. Their influence reveals the developing mathematical and scientific interests of Cracow masters.

325. Marsilius of Inghen. Questiones subillissime Johannis Marcilii Inguen super octo libros phisicorum secundum nominalium viam. Lyon, 1518. Reprint. Frankfort: Minerva, 1964.

This Physics commentary expresses views not usually identified with Marsilius.

326. Michalski, Konstanty. La philosophie au XIVe siècle: Six études. Edited by Kurt Flasch. Frankfort: Minerva, 1969. Pp. xv + 413.

A collection of six essays originally published during 1922-1937 on currents in fourteenth-century theology, logic and natural philosophy illustrated in the thought of not only the major figures but also by the works of less well-known masters.

327. Molland, A.G. "The Geometrical Background to the 'Merton School:' An Exploration into the Application of Mathematics to Natural Philosophy in the Fourteenth Century." British Journal for the History of Science, 4 (1968-69), 108-125.

Argues that scholastics displayed little creativity in the realm of pure geometry, but in the application of geometry to natural philosophy, for example, in the fourteenth-century applications of proportionality theory to nature or to problems ad imaginationem, scholastics revealed considerable ingenuity.

328. Moody, Ernest A. "Galileo and his Precursors." Galileo Reappraised. Edited by Carlo L. Golino. Berkeley: University of California Press, 1966, pp. 23-43.

Argues that Galileo's mature mechanics was influenced by ideas developed at fourteenth century Paris and Oxford such as the permanent impetus of Buridan, the kinematic analysis of motion of the Merton scholars and the graphical representation of qualitative intension and remission created by Oresme. However, it was Galileo who constructed a new science of mechanics from these disparate and isolated elements.

329. Moody, Ernest A. "Laws of Motion in Medieval Physics." Studies in Medieval Philosophy, Science, and Logic: Collected Papers 1933-1969. Berkeley: University of California Press, 1975, pp. 189-201.

Sketches developments in dynamics and kinematics at fourteenth-century Oxford and Paris which marked the beginnings of notions that went into classical physics.

330. Moody, Ernest A. "Ockham and Aegidius of Rome." Studies in Medieval Philosophy, Science, and Logic: Collected Papers, 1933-1969. Berkeley: University of California Press, 1975, pp. 161-188.

Argues that extracts from Aegidius in Ockham's Expositio on the Physics on Text 71 reveal Aegidius as a supporter of the doctrine that quantity is an absolute thing distinct from substances and qualities. On this point, as well as others, Aegidius was a primary opponent of Ockham.

331. Moser, Simon. Grundbegriffe der Naturphilosophie bei Wilhelm von Ockham. Innsbruck: Felizian Rauch, 1932. Pp. vi + 176.

Compares several of the fundamental concepts of Ockham's thought in his Summulae in libros physicorum with the opinions in Aristotle, including form, matter and privation, the four causes, motion and time.

332. Murdoch, John E. Rationes mathematice: Un aspect du rapport des mathématiques et la philosophie au moyen âge. Paris: Palais de la Découverte, 1962. Pp. 37.

Maintains that while medieval applications of mathematics to problems in kinematics and dynamics are familiar, the employment of mathematical arguments in issues of more philosophical import is less well-known. Illustrates this use of mathematical reasoning as applied to three areas: (1) the validity of astrological prediction; (2) the possibility of the eternity of the world; and, (3) the composition of the continuum.

333. Murdoch, John E. "Mathesis in philosophiam scholasticam introducta: The Rise and Development of the

Application of Mathematics in Fourteenth-Century Philosophy and Theology." Arts libéraux et philosophie au moyen âge. Actes du quatrième congrès international de philosophie médiévale. Montreal: Institut d'etudes médiévales, 1969, pp. 215-254.

Illustrates the applications of mathematics to fourteenth-century natural philosophy and theology as seen in controversies about atomism, the eternity of the world, the intension and remission of qualities and the perfection of species.

334. Murdoch, John E. "Naissance et développement de l'atomisme au bas moyen âge latin." La science de la nature: Théories et pratiques. Cahiers d'études médiévales, II. Paris: J. Vrin, 1974, pp. 11-32.

Shows that medieval atomism, an independent creation in the early fourteenth century, in the work of Henry of Harclay, Walter Chatton, Gerard of Odon and Nicolas Bonettus, was not devised to explain the actual structure of nature or of physical processes, but was rather employed to explain space, time, motion in an abstract and mathematical fashion.

335. Murdoch, John E. "Music and Natural Philosophy: Hitherto Unnoticed Questiones by Blasius of Parma (?)." Science, Medicine and the University: 1200-1550. Essays in Honor of Pearl Kibre, I. Edited by Nancy G. Siraisi and Luke Demaitre. Manuscripta, 20 (1976), pp. 119-136.

Argues that the doctrinal similarities and references to topics treated in Blasius' works, as well as the wide range of subject matter in addition to music, especially the "calculatory" methods (which Blasius knew well), attest to his likely authorship.

336. Murdoch, John E. "Subtilitates Anglicanae in Fourteenth-Century Paris: John of Mirecourt and Peter Ceffons." Machaut's World: Science and Art in the Fourteenth Century. Edited by Madeleine Pelner Cosman and Bruce Chandler. New York: New York Academy of Sciences, 1978, pp. 51-86.

Examines the impact of four aspects of the Oxford "calculatory tradition" on theological thought at Paris as seen in the commentaries on the Sentences of Peter Lombard of these two relatively little-known Cistercians. While the transmission to and elaboration of Oxford ideas on Parisian natural philosophy is well-known, their application to theological themes has not been explored.

337. Murdoch, John E. "Propositional Analysis in Fourteenth Century Natural Philosophy: A Case Study." Synthèse, 40 (1979), 117-146.

Argues that fourteenth-century scholastics treated problems dealing with the extrinsic or intrinsic limits of continuous change propositionally or metalinguistically which introduced a tradition of logical problems concerned with the terms incipit-desinit. These problems were used in the construction of sophisms, for example, by Heytesbury.

338. Murdoch, John E. "Henry of Harclay and the Infinite." Studi sul XIV secolo in memoria Anneliese Maier. Edited by Alfonso Maierù and Agostino Paravicini Bagliani. Rome: Edizioni di storia e letteratura. 1981, pp. 219-261.

Analyzes two questions by Henry who held non-Aristotelian views on infinity and the continuous--that one infinite can be greater than another, that continua are composed of an infinite number of indivisibles, each next to the other, and that an actual infinite is possible. The questions discussed here emphasize the inequality of infinites and the infinite number of indivisibles in continua.

339. Murdoch, John E. "Scientia mediantibus vocibus: Metalinguistic Analysis in Late Medieval Natural Philosophy." Sprache und Erkenntnis im Mittelalter: Akten des VI. internationalen Kongresses für mittelalterliche Philosophie, I. Berlin: de Gruyter, 1981, pp. 73-106.

Traces the background and development of the use of metalinguistic analysis in fourteenth century natural philosophy, that is, the treatment of problems concerning nature within the framework of propositions and their terms, and describes metalinguistic analysis as applied to two aspects of medieval discussions of indivisibles and their place in continua--the treatment of indivisibles as limits of continua and the existence of indivisibles.

340. Murdoch, John E. "William of Ockham and the Logic of Infinity and Continuity." Infinity and Continuity in Ancient and Medieval Thought. Edited by Norman Kretzmann. Ithaca: Cornell University Press, 1982, pp. 165-206.

Discusses three aspects of the problem of infinity and the continuous which particularly interested Ockham--whether infinites are equal or not, the existence of indivisibles and their role in natural philosophy, and the infinite divisibility of continua.

341. Murdoch, John E. and E. Synan. "Two Questions on the Continuum: Walter Chatton (?) O.F.M. and Adam Wodeham, O.F.M." Franciscan Studies, n. s., 26 (1966), 212-288.

Maintains that these two questions by Chatton, who held the position that continua are composed of indivisibles, and Wodeham, who defended the almost universally accepted Aristotelian view of continuous magnitude, reveal the English role in this controversy and supply evidence of the positions of other English scholastics. Wodeham's question, drawn from his Tractatus de indivisibilibus, specifically names Henry of Harclay, the other member of the English branch of fourteenth-century atomism, and shows his familiarity with William of Alnwick's refutation of Harclay. The pro-atomist question, represented by Chatton, is probably by him as it contains opinions expressed in his Sentences commentary.

342. Ockham, William of. De sacramento altaris. Edited by T. Bruce Birch. Burlington, Ind.: Lutheran Literary Board, 1930. Pp. xlvii + [1] + 567.

Probably only the second treatise in this edition should be called De sacramento altaris, although the first is also on the Eucharist. See item 311.

343. Ockham, William of. Tractatus de successivis. Edited by Philotheus Boehner. St. Bonaventure, N.Y.: Franciscan Institute, 1944. Pp. 122.

344. Ockham, William of. "Questioni inedite di Occam sul tempo." Edited by Francesco Corvino. Rivista critica di storia della filosofia, 11 (1956), 41-67; 12 (1957), 42-63.

An edition of the questions (35-57) in Ockham's *Questiones in libros physicorum*, Book IV, dealing with time.

345. Ockham, William of. *Philosophical Writings*. Edited and translated by Philotheus Boehner. New York: Nelson, 1957. Pp. lix + 154.

Selections from Ockham's major works dealing with epistemology, logic, theology, natural philosophy and ethics.

346. Ockham, William of. "Questioni inedite sul continuo." Edited by Francesco Corvino. *Rivista critica di storia della filosofia*, 13 (1958), 191-208.

An edition of questions from the *Questiones in libros physicorum* on the nature of the continuum.

347. Ockham, William of. *Quotlibeta septem and Tractatus de sacramento altaris*. Strasbourg, 1491. Reprint. Louvain: Editions de la Bibliothèque S.J., 1962.

348. Ockham, William of. *Philosophia naturalis Guilielmi Occham*. Rome, 1637. Reprint. London: Gregg Press, 1963.

This work is also known as the *Summulae in libros Physicorum*.

349. Ockham, William of. *Venerabilis inceptoris Guillelmi de Ockham Quodlibeta septem*. Edited by Joseph C. Wey. St. Bonaventure, N.Y.: St. Bonaventure University, 1980. Pp. 41, 838.

350. O'Donnell, J.R. "Nicholas of Autrecourt." *Mediaeval Studies*, 1 (1939), 179-280.

Contains the text of Autrecourt's *Exit ordo executionis*.

351. Oresme, Nicole. "*Inter omnes impressiones*." Edited by René Mathieu. *Archives d'histoire doctrinale et littéraire du moyen âge*, 26 (1959), 277-294.

Identifies this work as by Oresme; however, Clagett questions this attribution. See item 35, p. 645.

352. Oresme, Nicole. Le livre du ciel et du monde. Edited by Albert D. Menut and Alexander J. Denomy. Translated with an introduction by Albert D. Menut. Madison: University of Wisconsin Press, 1968. Pp. xiii + 778.

Represents the first Aristotelian scientific treatise with commentary translated into a vernacular. Written in 1377, it was among the Aristotelian works translated into French by Oresme for Charles V.

353. Oresme, Nicole. Nicole Oresme and the Medieval Geometry of Qualities and Motions: A Treatise on the Uniformity and Difformity of Intensities Known as Tractatus de configurationibus qualitatum et motuum. Edited with an introduction, English translation, and commentary by Marshall Clagett. Madison: University of Wisconsin Press, 1968. Pp. xiii + 713.

An edition of Oresme's configuration doctrine treatise.

354. Paul of Venice. Summa philosophiae naturalis. Venice, 1503. Reprint. Hildesheim: Olms, 1974.

355. Shapiro, Herman. "Motion, Time, and Place According to William Ockham." Franciscan Studies, 16 (1956), 213-303, 319-372.

Analyzes Ockham's arguments for his own views on the nature of motion, time and space, and his refutations of the arguments of his opponents as developed primarily in the Philosophia naturalis (Summulae in libros physicorum). Ockham's purpose was to show that motion (defined broadly to include mutation), time and space were not absolute entities apart from individual substances and qualities.

356. Shapiro, Herman. "Walter Burley and the Intension and Remission of Forms." Speculum, 34 (1959), 413-427.

Points out that Burley rejected both the addition and admixture theories of the intension and remission of forms. In his interpretation, a form or part of a form was, e.g., acquired in intension, but the new degree of the quality was not added to those already existing, rather it was completely new, the previous form having been destroyed.

357. Shapiro, Herman. "Walter Burley and Text 71." Traditio, 16 (1960), 395-404.

Shows that Burley was one fourteenth-century scholastic who attempted to reconcile the two opposed positions on the essential requirement for the occurrence of motion arising from Averroes' commentary on Physics IV, 8 (known as Text 71). While Averroes had maintained with Aristotle that motion requires a resistant medium, he reported the opinion of Avempace that the time of the mobile's transversal of empty space was the essential factor. Burley's compromise redefined the concept of void space to which he attributed a positive resistant value.

358. Skabelund, Donald and Phillip Thomas. "Walter of Odington's Mathematical Treatment of the Primary Qualities." Isis, 60 (1969), 331-350.

Points out that in the final chapter of his Icocedron, Walter provides rules for determining the resultant intensity in compositions of the four primary qualities, taken two at a time, and uses an exponential function stemming from Alkindi employed by Arnald of Villanova and Bradwardine. See item 1382.

359. Stamm, Edward. "Tractatus de continuo von Thomas Bradwardine." Isis, 26 (1936), 13-32.

Outlines the content of this treatise in which Bradwardine defends the infinite divisibility of continua.

360. Stump, Eleanore. "Theology and Physics in De sacramento altaris: Ockham's Theory of Indivisibles." Infinity and Continuity in Ancient and Medieval Thought. Edited by Norman Kretzmann. Ithaca: Cornell University Press, 1982, pp. 207-230.

Points out that Ockham's account of indivisibles and his treatment of quantity form part of his discussion of the Eucharist. To Ockham, neither were absolute, separately existing things. Indivisibles are privations, limits of continua. But Ockham's theory of language provided a way to allow indivisibles to play a role in mathematics and natural philosophy.

361. Sylla, Edith D. "Medieval Quantifications of Qualities: The 'Merton School'." Archive for History of Exact Sciences, 8 (1971), 7-39.

Discusses developments in the quantification of qualities in the work of Dumbleton, Heytesbury, Swineshead and Bradwardine prior to Oresme's geometrical treatment. These Mertonians drew on the generally accepted theory of intension and remission by addition which opened the way for a quantified treatment of qualities founded on intensities.

362. Sylla, Edith D. "Medieval Concepts of the Latitude of Forms: The Oxford Calculators." Archives d'histoire doctrinale et littéraire du moyen âge, 40 (1973), 223-283.

Discusses the very different treatments of the latitude doctrine in the thought of three Oxford authors (Walter Burley, Roger Swineshead, John Dumbleton), and compares them with the subsequent configuration doctrine of Oresme.

363. Sylla, Edith D. "Autonomous and Handmaiden Science: St. Thomas and William of Ockham on the Physics of the Eucharist." The Cultural Context of Medieval Learning. Edited by John E. Murdoch and Edith D. Sylla. Dordrecht: Reidel, 1975, pp. 349-396.

Illustrates the differences scholastics could reveal in their conception of the relationship between theology and natural philosophy by the opinions of Aquinas and Ockham on the Eucharist, a topic which involved philosophical issues such as the relation of substances and qualities and the nature of quantity. Aquinas was prepared to modify natural philosophy so as to subordinate it to theological doctrine, while Ockham permitted philosophy an autonomous role, adducing miraculous intervention when required.

364. Sylla, Edith D. "Godfrey of Fontaines on Motion with Respect to Quantity of the Eucharist." Studi sul XIV secolo in memoria Anneliese Maier. Edited by Alfonso Maierù and Agostino Paravicini Bagliani. Rome: Edizioni di storia e letteratura, 1981, pp. 105-141.

Points out that important views in scholastic natural philosophy often arose in the context of theological problems. Here, given medieval (Aristotelian) notions as to quantity and change in quantity, how was this to the understood in the case of the Eucharist. Various solutions were proposed: Aegidius Romanus relied on the concept of indeterminate dimensions which Godfrey in the quodlibetal questions treated here, rejected in favor of a succession of quantities (forms) theory.

365. Sylla, Edith D. "Infinite Indivisibles and Continuity in Fourteenth-Century Theories of Alteration." _Infinity and Continuity in Ancient and Medieval Thought_. Edited by Norman Kretzmann. Ithaca: Cornell University Press, 1982, pp. 231-257.

Points out that fourteenth-century scholastics (notably Burley) who adhered to a succession of forms explanation of alteration (rather than the addition theory), which required an infinite succession of indivisible degrees, had to reconcile these indivisibles with the continuity of motion. Explores the resolution of this problem as seen in texts of Walter Burley and Richard Kilvington. Both scholastics relied on available features of logic to effect their solutions.

366. Wallace, William A. "Mechanics from Bradwardine to Galileo." _Journal of the History of Ideas_, 32 (1971), 15-28.

Argues that while a quantitative treatment of motion, a salient aspect of classical mechanics, clearly began with the Merton _calculatores_, it is likewise possible to trace the gradual development of a change in outlook in scholastic discussions from the fourteenth through the sixteenth centuries which prepared the way for the adoption of experimental treatments of motion. This development was perceptible in an increasing focus less on abstract knematics than on problems in dynamics related to the motions of actual physical bodies.

367. Weinberg, Julius R. _Nicolaus of Autrecourt_: _A Study in 14th Century Thought_. Princeton, 1948. Reprint. New York: Greenwood Press, 1969. Pp. 242.

Argues that Autrecourt's severe limitation of certainty led to a critique of the possibility of natural knowledge and of the Aristotelian natural philosophy espoused by his contemporaries. As this philosophy and its structure of the world could not be demonstrated, another could be shown as more probable, such as the existence of atoms, the void and the eternity of the world.

368. Weisheipl, James A. "The Place of John Dumbleton in the Merton School." Isis, 50 (1959), 439-454.

Characterizes the thought of the Mertonian Dumbleton revealed in his Summa logicae et philosophiae as dependent on both Ockham's nominalism and the mathematical approach to physical problems initiated by Bradwardine. Concludes that Dumbleton seems unaware of the philosophical discrepancies between his sources, and that his version of Bradwardine's methods is not too satisfactory.

369. Weisheipl, James A. "The Concept of Matter in Fourteenth Century Science." The Concept of Matter in Greek and Medieval Philosophy. Edited by Ernan McMullin. Notre Dame: University of Notre Dame Press, 1963, pp. 147-169.

Argues that while thirteenth-century scholastics, e.g., Albertus Magnus and Aquinas, treated problems arising from the quantitas materiae largely within an Aristotelian framework, the discussions of fourteenth century schoolmen, e.g., Ockham and the Oxford calculatores, differed widely from one another and transcended the original Aristotelian source.

370. Weisheipl, James A. "Roger Swyneshed, O.S.B., Logician, Natural Philosopher and Theologian." Oxford Studies Presented to Daniel Callus. Oxford: Clarendon Press for the Oxford Historical Society, 1964, pp. 231-252.

Disentangles the three English scholastics surnamed "Swyneshed." John and Richard were fellows at Merton, Roger, senior to them, was at Oxford, became a master of theology and spent his final years as a Benedictine monk at Glastonbury. John,who was a canonist, seems to have written nothing which has survived, while Richard was the author of the Liber calculationum. Roger wrote logical works and a De motus naturalibus, all in the early 1330's.

371. Weisheipl, James A. "Ockham and Some Mertonians." Mediaeval Studies, 30 (1968), 163-213.

Traces the careers, and discusses the work of fourteenth-century Oxford masters who were at Merton College in connection with their response to Ockham's doctrines. Both Walter Burley and Bradwardine (the founder of the Merton school which stressed the application of mathematics to natural philosophy) held views antithetical to Ockham. Heytesbury was an intellectual disciple of both Bradwardine and Ockham, while Dumbleton accepted nominalist logic and Ockham's natural philosophy in a more overt manner. Richard Swineshead succeeded in separating the mathematical analysis of physical problems from Ockham's natural philosophy.

372. Wieleitner, H. "Der Tractatus de latitudinibus formarum des Oresme." Biblioteca mathematica, ser. 3, 13 (1912-13), 115-145.

This is actually a partial edition of a work by Jacobus de Sancto Martino, probably written around 1390, based on Oresme's configuration doctrine but containing some differences.

373. Wilson, Curtis. William Heytesbury: Medieval Logic and the Rise of Mathematical Physics. Madison: University of Wisconsin Press, 1960. Pp. xii + 219.

Analyzes the mathematical and physical chapters of the Regule solvendi sophismata written in 1335. The work is characteristic of the particularly medieval combination of mathematics, logic and physics.

374. Zoubov, V.P. "Walter Catton, Gerard d'Odon et Nicolas Bonet." Physis, 1 (1959), 261-278.

Discusses the works and ideas of these three proponents of the view that continua are composed of indivisibles, their relationship to one another and the attacks of their chief critics. While their arguments for indivisibles had philosophical aspects, the major field of contention lay in geometry since they were forced to defend themselves against the geometrical criticisms proposed by their opponents.

375. Zoubov, V.P. "Jean Buridan et les concepts du point au quatorzième siècle." Mediaeval and Renaissance Studies, 5 (1961), 42-95.

Presents the text of Buridan's De puncto in which the Parisian master defends the position that a point has no real positive existence distinct from a line against a "magister Montescalesio."

376. Zoubov, V.P. "Nicole Oresme et la musique." Mediaeval and Renaissance Studies, 5 (1961), 96-107.

Argues that Oresme's interest in music is attested by the frequent references in his treatises as well as his composition of a work on the monochord. His aesthetic sense in music, as in astronomy, stressed variety as superior to a monotonous uniformity.

377. Zoubov, V.P. "Blaise de Parme et la principe d'inertie." Physis, 5 (1963), 39-42.

Argues that Maier's claim that Blasius of Parma formulated the principle of inertia in his questions on the Physics is not supported by the text. As is apparent in another passage, Blasius accounted for projectile motion as due either to the surrrounding air or to impetus. See item 206, p. 143.

## OPTICS

378. Alhazen. (Ibn al-Haytham). "Ibn al-Haitams Schrift über parabolische Hohlspiegel." Edited by J.L. Heiberg and Eilhard Wiedemann. Biblioteca mathematica, ser. 3, 10, (1910), 201-237.

A German translation based on Arabic and Latin texts of Alhazen's On Burning Mirrors.

379. Alhazen (Ibn al-Haytham). De aspectibus. Opticae thesaurus Alhazeni Arabis libri septem .... Edited by Friedrich Risner. Basel, 1572. Reprint, with introduction by David C. Lindberg. New York: Johnson Reprint, 1972.

380. Björnbo, A.A. and Sebastian Vogl, eds. Alkindi, Tideus und pseudo-Euklid: Drei optische Werke. Abhandlungen zur Geschichte der mathematischen Wissenschaften mit Einschluss ihrer Anwendungen, 26, 3. Leipzig: Teubner, 1912. Pp. vi + 176.

Optical works translated by Gerard of Cremona including Alkindi, De aspectibus, Tideus, De speculis, and pseudo-Euclid, De speculis.

381. Blasius of Parma. "Le Questioni di perspectiva di Biago Pelacani da Parma." Edited by Graziella Federici Vescovini. Rinascimento, 2d ser., 12 (1961), 163-250.

Edition and analysis of the Questiones perspectivae, Book I, QQ 14 and 16, and the question on the rainbow from Book III.

382. Blasius of Parma. "Questioni inedite di ottica di Biagio Pelacani da Parma." Edited by Franco Alessio. Rivista critica di storia della filosofia, 16 (1961), 79-110; 188-221.

An edition of Book I, QQ 1-10 of Blasius' Questiones perspectivae. See item 381.

383. Boyer, Carl. "Robert Grosseteste on the Rainbow." Osiris, 11 (1954), 247-258.

Maintains that Grosseteste was the first to connect the bow with refraction.

384. Boyer, Carl B. "The Theory of the Rainbow: Medieval Triumph and Failure." Isis, 49 (1958), 378-390.

Argues that while the Dominican Theodoric of Freiberg did give a correct explanation of the rainbow as due to two refractions and a reflection in the individual raindrop, he left much unexplained.

385. Eastwood, Bruce S. "Robert Grosseteste's Theory of the Rainbow: A Chapter in the History of Non-experimental Science." Archives internationales d'histoire des sciences, 19 (1966), 313-332.

Argues that Grosseteste's introduction of refraction into his theory of the rainbow (previous theories had relied on reflection only), was not based on experimental procedures as was the correct explanation employing both refraction and reflection developed later by Theodoric of Freiberg. Grosseteste had a scientific methodology which included the experimental verification or falsification of a theory, but in his work on the rainbow, he relied on reports of authorities and metaphysical principles of simplicity and economy.

386. Eastwood, Bruce S. "Grosseteste's 'Quantitative' Law of Refraction: A Chapter in the History of Non-experimental Science." Journal of the History of Ideas, 28 (1967), 403-414.

Argues that Grosseteste's peculiar law, stating that the angle of refraction is equal to half the incident angle, was not based on his own acknowledged experimental method, but rather on philosophical and metaphysical *a priori* convictions. Guided by principles of uniformity and economy, he concluded that as a relationship of equality of angles governs the law of reflection, this feature of equality must likewise be true for refraction.

387. Eastwood, Bruce S. "Averroes' View of the Retina--A Reappraisal." Journal of the History of Medicine and Allied Sciences, 24 (1969), 77-82.

Establishes that Averroes located photoreceptivity at the lens of the eye rather than at the retina.

388. Eastwood, Bruce S. "Metaphysical Derivations of a Law of Refraction: Damianos and Grosseteste." Archive for History of Exact Sciences, 6 (1970), 224-236.

Points out that both Damianos (4th century A.D.) and Grosseteste arrived at the same erroneous law of refraction (angle of refraction is one-half the incident angle), based on the same metaphysical presuppositions, the principles of simplicity and uniformity. Damianos was not known in the medieval Latin West, but both he and Grosseteste were subject to the same Neoplatonic influences and both believed metaphysical principles more important than authority or experience.

389. Eastwood, Bruce S. "al-Fārābī on Extramission, Intromission and the Use of Platonic Visual Theory." Isis, 70 (1979), 423-425.

Argues that al-Fārābī's notions of vision should be set in the context of his synthesis of Aristotelian and Platonic ideas.

390. Eastwood, Bruce S. The Elements of Vision: The Micro-Cosmology of Galenic Visual Theory According to Hunayn ibn Ishāq. Transactions of the American Philosophical Society, 72, part 5. Philadelphia: American Philosophical Society, 1982. Pp. 58.

Argues from an analysis of three tracts of Hunayn's Ten Treatises on the Eye that his anatomy and physiology of the eye was founded on cosmological principles applied to the eye as a microcosm. This notion, Galenic in origin, was developed in the context of medical practice.

391. Euclid. "Liber de visu: The Greco-Latin Translation of Euclid's Optics." Edited by Wilfred R. Theisen. Mediaeval Studies, 41 (1979), 44-105.

Points out in the introduction to the edition the important position held by the work among ancient and Moslem authors and its enthusiastic reception in the Latin West. It presents Euclid's theory of vision, based on visual rays conceived in geometrical terms. Reaching Europe before the end of the twelfth century, it was also translated more than once from Arabic.

392. Federici Vescovini, Graziella. "Les Questions de perspectiva de Dominicus de Clivaxo." Centaurus, 10 (1964), 14-28.

Describes the Questiones de perspectiva, at present extant in a unique MS at the National Library at Florence. The first and last of its six questions are edited here.

393. Federici Vescovini, Graziella. Studi sulla prospettiva medievale. Turin: Giappichelli, 1965. Pp. 286.

A series of essays, chronologically ordered, which treats the science of perspective in broad terms, including theological, ontological and epistemological implications as well as medical, mathematical and experimental aspects.

394. Grosseteste, Robert. <u>On Light (De luce)</u>. Translated by Clare C. Riedl. Milwaukee: Marquette University Press, 1942. Pp. 17.

395. Koelbing, Huldrych M. "Averroes' Concepts of Ocular Function--Another View." <u>Journal of the History of Medicine and Allied Sciences</u>, 27 (1972), 207-213.

Argues, based on Averroes' <u>Epitome</u> of the <u>Parva naturalia</u> rather than on his <u>Colliget</u>, that Averroes concluded that sensitivity is the property of the retina, not the lens of the eye, counter to Eastwood's opinion. See items 387, 406.

396. Lindberg, David C. "The <u>Perspectiva communis</u> of John Pecham: Its Influences, Sources, and Content." <u>Archives internationales d'histoire des sciences</u>, 18 (1966), 37-53.

Maintains that Pecham's work was conceived as an introduction to Alhazen's <u>Perspectiva</u>. Pecham accepted Alhazen's intromission theory of vision, quoted him extensively and structured his own treatise after that of Alhazen.

397. Lindberg, David C. "Roger Bacon's Theory of the Rainbow: Progress or Regress?" <u>Isis</u>, 57 (1966), 235-248.

Argues that since the correct explanation of the rainbow does require refraction, which Grosseteste had been the first to introduce in his own theory of the bow, the rejection of refraction in favor of reflection by Bacon a generation later might be considered retrogressive. However, Grosseteste's theory did not employ refraction properly and did not show how it could account for the bow. Within the context of Grosseteste's inadequate refraction theory, Bacon's rejection was thus not a backward step. Bacon's own account is more reasonable, and he did focus on the individual raindrop which Grosseteste had not done.

398. Lindberg, David C. "Alhazen's Theory of Vision and its Reception in the West." <u>Isis</u>, 58 (1967), 321-341.

Argues that Alhazen's Perspectiva, translated in the late twelfth or early thirteenth century, provided a comprehensive theory of vision which denied visual rays. Objects, whether self-luminous or illuminated, were seen when rectilinear rays of light passed from every point in the visual field to the lens of the eye which was the sensitive organ of vision as Galen had maintained. The base of the visual cone is thus on the object with its apex on the surface of the lens. Alhazen's work provided the western optical tradition with an intromission theory of vision which was incorporated in the optical works of Bacon, Pecham and Witelo.

399. Lindberg, David C. "The Cause of Refraction in Medieval Optics." British Journal for the History of Science, 4 (1968), 23-38.

Stresses that ancient and medieval optical writers distinguished between a determination of the causes of refraction and the determination of a quantitative law of refraction. Following a hint by Ptolemy that the cause was related to the densities of the media, Alhazen, Bacon, Pecham and Witelo assumed that the cause was the resistance offered to light by media of different densities and the effect of this on its motion. This mechanistic approach continued into the seventeenth century.

400. Lindberg, David C. "The Theory of Pinhole Images from Antiquity to the Thirteenth Century." Archive for History of Exact Sciences, 5 (1968), 154-176.

Traces the history of pinhole images formed by finite-sized apertures in the pseudo-Aristotelian Problemata, Alkindi's De aspectibus, in the pseudo-Euclidean De speculis and in the works of thirteenth-century optical writers, Roger Bacon, Witelo and Pecham. The problem these authors did not solve was how to explain that a noncircular opening gave rise to a circular image assuming the rectilinear propagation of light.

401. Lindberg, David C. "A Reconsideration of Roger Bacon's Theory of Pinhole Images." Archive for History of Exact Sciences, 6 (1970), 214-223.

Argues that thanks to a previously unconsidered treatise attributed to Bacon, the De speculis comburentis, which deals extensively with pinhole images, Bacon's role in the medieval treatment of the theory and his influence on Pecham must be re-evaluated.

402. Lindberg, David C. "The Theory of Pinhole Images in the Fourteenth Century." Archive for History of Exact Sciences, 6 (1970), 299-325.

Treats pinhole images in the optical works of Henry of Langenstein, Blasius of Parma, and in discussions by Egidius of Baisu and Levi ben Gerson. The theory in the fourteenth century was an unsolved legacy provided by the Perspectiva communis of Pecham. Concludes that this century was further from the correct theory than was the preceding one.

403. Lindberg, David C. "Alkindi's Critique of Euclid's Theory of Vision." Isis, 62 (1971), 469-489.

Argues that Alkindi, in his De aspectibus, although he followed the emission theory of vision adopted by Euclid, gave the latter's largely geometrical presentation as more physical content.

404. Lindberg, David C. "Bacon, Witelo and Pecham: The Problem of Influence." Actes du XIIe congrès international d'histoire des sciences, IIIA. Paris: Blanchard, 1971, pp. 103-107.

Shows that Bacon's optical works were undoubtedly known to Pecham, and that it is highly likely that Witelo read them during his extended stay at Viterbo. Although the optical treatises of Pecham and Witelo were written at approximately the same time, it is more probable that Pecham borrowed from Witelo than vice versa.

405. Lindberg, David C. "Lines of Influence in Thirteenth-Century Optics: Bacon, Witelo, Pecham." Speculum, 46 (1971), 66-83.

Discusses problems of not only who influenced whom, but by what means that influence was exercised within the broader framework of the human and social aspects of medieval science.

406. Lindberg, David C. "Did Averroes Discover Retinal Sensitivity?" Bulletin of the History of Medicine, 49 (1975), 273-278.

Rejects the opinion that Averroes held that the retina was the light-sensitive organ of the eye. Passages in the Colliget and the Epitome of the Parva naturalia reveal that, while the retina played an important role in sight, the crystalline lens was the principal organ of vision; this opinion was also held by Galen. See item 395.

407. Lindberg, David C. "The Intromission-Extramission Controversy in Islamic Visual Theory: al-Kindī versus Avicenna." Studies in Perception. Edited by P.K. Machamer and R.G. Turnbull. Columbus: Ohio State University Press, 1978, pp. 137-159.

Argues that this controversy involved factors more complex than just the direction of radiation. Alkindi argued against intromission from a mathematical viewpoint, while Avicenna rejected extramission for physical reasons. These authors reflect two traditions in medieval optics akin to the dilemma confronting medieval astronomers; the two positions were reconciled in Alhazen.

408. Lindberg, David C. "Medieval Latin Theories of the Speed of Light." Roemer et la vitesse de la lumière. Paris: J. Vrin, 1978, pp. 45-72.

Argues that the medieval treatment of the speed of light was connected with notions of how time was involved in visual perception; among those who thought the propagation of light occurred in time, its speed was considered imperceptible. Traces opinions from Augustine and the Platonic tradition, Aristotle and Alhazen, the optical writers of the thirteenth century (Bacon and Witelo) and from the fourteenth-century scholastics Buridan and Blasius of Parma.

409. Marshall, Peter. "Nicole Oresme on the Nature, Reflection, and Speed of Light." Isis, 72 (1981), 357-374.

Traces the Aristotelian and perspectivist traditions in Oresme's theories on light in his Questions on De anima.

410. Pecham, John. John Pecham and the Science of Optics: Perspectiva communis. Edited with an introduction, English translation and critical notes by David C. Lindberg. Madison: University of Wisconsin Press, 1970. Pp. xvii + 300.

411. Pecham, John. Tractatus de perspectiva. Edited with an introduction and notes by David C. Lindberg. St. Bonaventure, N.Y.: Franciscan Institute, 1972. Pp. 110.

412. Sabra, A.I. "The Authorship of the Liber de crepusculis, an Eleventh Century Work on Atmospheric Refraction." Isis, 58 (1967), 77-58.

Shows this work dealing with atmospheric refraction usually attributed to Alhazen, was probably written by Abu ibn Mu'adh, an Andalusian mathematician of the eleventh century.

413. Sabra, A.I. "The Physical and the Mathematical in Ibn al-Haytham's Theory of Light and Vision." The Commemoration Volume of Bīrūnī International Congress in Tehran. B. English and French Papers. Tehran: High Council of Culture and Art, 1976, pp. 439-478.

Discusses Alhazen's deliberate synthesis of the optical doctrines of the physicists (forms come to the eye from objects) and the mathematicians (doctrine of visual rays). His own theory of vision combined features of both these traditions with the notion of idola from the Greek atomists.

414. Theisen, Wilfred. "Witelo's Recension of Euclid's De visu." Traditio, 33 (1977), 394-402.

Establishes that Witelo was not only familiar with Euclid's De visu, but made a recension of the work prior to writing his Perspectiva. Although the recension is not generally attributed to Witelo, the many similarities between that treatise and the Perspectiva attest that Witelo used it in the composition of the latter.

415. Theisen, Wilfred. "A Note on John of Beaumont's Version of Euclid's De visu." British Journal for the History of Science, 11 (1978), 151-155.

Stresses the enthusiastic reception of Euclid's optical treatise in the Latin West, available in translation from Arabic and Greek before the end of the twelfth century. Focuses on the commentary owing to a John of Beaumont. While only an introduction, the presence of this version and others in various manuscript collections attests to wide diffusion in the MSS.

416. Theodoric of Freiberg. De iride et radialibus impressionibus. Edited by Joseph Würschmidt. Beiträge zur Geschichte der Philosophie des Mittelalters, 12. Münster: Aschendorff, 1914. Pp. ix + 205.

An edition of Theodoric's work on the rainbow.

417. Wiedmann, Eilhard. "Über den Apparat zur Untersuchung und Brechung des Lichtes von Ibn al-Haitam." Annalen der Physik und Chemie, n.s., 21 (1884), 541-544.

German translation of a portion of the passage from Alhazen's optical treatise dealing with his instrument for the investigation of refraction.

418. Witelo. Perspectiva. Opticae thesaurus Alhazeni Arabis libri septem ... Item Vitellonis Thuringopoloni libri X. Edited by Friedrich Risner. Basel, 1572. Reprint with introduction by David C. Lindberg. New York: Johnson Reprint, 1972.

419. Witelo. Witelonis Perspectivae Liber primus. Book I of Witelo's Perspectiva. Edited and translated with introduction and commentary by Sabetai Unguru. Studia Copernicana, 15. Warsaw: Polish Academy of Sciences, 1977. Pp. 330.

## ASTRONOMY: COMPUTI

420. Bede. Bedae Opera de temporibus. Edited by C.W. Jones. Cambridge, Mass.: Mediaeval Academy of America, 1943. Pp. xiii + 415.

Traces the history of Easter cycles and of the computus genre, and presents editions of Bede's two computistical treatises--the De temporibus (703) and the less condensed De temporum ratione (725).

421. Cordoliani, Alfred. "Le comput de Dicuil." Cahiers de civilisation médiévale, 3 (1960), 325-337.

Describes the Liber de astronomia et computo of this Irish scholar composed during 814/816 and dedicated to Louis the Pious. Points out that the treatise has a fifth book in two chapters (on the structure of the Easter cycle and on epacts) in addition to the four previously edited.

422. Cordiolani, Alfred. "L'activité computistique de Robert, évêque de Hereford." Mélanges offerts à René Crozet, I. Edited by Pierre Gallais and Yves-Jean Riou. Poitiers: Société d'études médiévales, 1966, pp. 333-340.

Discusses Robert's computistical work in which he justifies the criticism of the eleventh-century chronicler, Marianus Scotus, who maintained that the currently accepted Dionysian era was twenty-two years too late.

423. Cordiolani, Alfred. "Le comput ecclésiastique à l'abbaye du Mont-Cassin au XIe siècle." Annuario di estudios medievales, 3 (1966), 65-89.

Discusses computistical texts from Monte Cassino which are based primarily on Bede.

424. Harrison, Kenneth P. "Luni-Solar Cycles: Their Accuracy and Usage." Saints, Scholars, and Heroes: Studies in Medieval Culture in Honour of Charles W. Jones, II. Edited by Margot H. King and Wesley M. Stevens. Collegeville, Minn.: St. Johns University Press, 1979, pp. 65-78.

Traces the fortunes of cycles relating the motions of the sun and moon between 200 and 800 A.D. from the pioneering work of Hippolytus of Rome who introduced an 8-year cycle through the 19-year cycle of Anatolius of Laodicea which eventually became standard in the Christian West.

425. Stevens, Wesley M. "Walahfrid Strabo--A Student at Fulda." Canadian Historical Association Historical Papers, 1971, pp. 13-20.

Discusses the problems Walahfrid faced in his study of Hraban's Computus at Reichenau, ca. 825. Walafrid corrected his copy of the text and added other scientific works to his personal notebook at Fulda until his death in 849.

426. Stevens, Wesley M. "Fulda Scribes at Work: Bodleian Library MS Canonici Miscellaneous 353." Bibliothek und Wissenschaft, 8 (1972), 287-316.

Analyzes the computus of Hrabanus Maurus, c. 836 A.D. See item 425.

427. Stevens, Wesley M. "Compotistica et astronomica in the Fulda School." Saints, Scholars, and Heroes: Studies in Medieval Culture in Honour of Charles W. Jones, II. Edited by Margot H. King and Wesley M. Stevens. Collegeville, Minn.: St. Johns University Press, 1979, pp. 27-64.

Argues that Carolingian scholars had interests in astronomy, arithmetic and geometry outside of the traditional framework of the liberal arts. This can be seen in the activities at Fulda of Hraban Maur and his pupils, Lupus of Ferrières and Walahfrid Strabo. The latter's personal notebook, now at St. Gall, is especially revealing as to how computistical studies were pursued in a monastic context in the ninth century.

## ASTRONOMY AND COSMOLOGY

428. Albategnius (al-Battānī). Opus astronomicum. 2 vols. Edited by C.A. Nallino. Milan, 1903-1907. Reprint. Frankfort: Minerva, 1969.

429. Alfonso X. Libros del saber de astronomia del rey D. Alfonso X de Castilla. 5 vols. Edited with notes and commentary by Manuel Rico y Sinobas. Madrid: Tip. de Don E. Aguado, 1863-1867.

Texts of astronomical-astrological works and treatises on astronomical instruments translated from Arabic to Castillian under the patronage of Alfonso the Wise. Vol. IV contains the Alfonsine Tables.

430. Alfraganus (al-Farġānī). Differentie in quibusdam collectis scientie astrorum. Edited by Francis J. Carmody. Berkeley: University of California Press, 1943. Pp. 60.

Latin text of the Elements of Astronomy in the translation of John of Seville.

431. Alpetragius (al-Biṭrūjī). De motibus celorum. Edited from the Latin translation of Michael Scot by Francis J. Carmody. Berkeley: University of California Press, 1952. Pp. 180.

An edition of al-Biṭrūjī's homocentric astronomy.

432. d'Alverny, Marie-Thérèse. "Survivances du 'système d'Héraclide' au moyen âge." Avant, avec, après Copernic: La representation de l'univers et ses conséquences épistémologiques. Paris: Blanchard, 1975, pp. 39-50.

Traces the treatment of the so-called system of Heraclides which placed the inferior planets Mercury and Venus in orbit around the sun in the works of early encyclopedists and commentators such as Pliny, Macrobius, Chalcidius, Martianus Capella as well as in John Scotus Erigena and Chartrian masters.

433. Aquinas, Thomas. Summa theologiae, vol. 10 (1a 65-74). Cosmogony. Edited and translated by William A. Wallace. New York: McGraw-Hill, 1967. Pp. xxiii + 255.

Has valuable appendixes on medieval science.

434. Bauer, Georg-Karl. Sternkunde und Sterndeutung der Deutschen in 9.-14. Jahrhundert. Berlin, 1937. Reprint. Nendeln/Liechtenstein: Kraus, 1967. Pp. 182.

Discusses astronomical knowledge among the German-speaking peoples, apart from technically-oriented and specialized astronomers, from the Carolingian

Renaissance to 1400, including notions of the cosmos, observations, connections with Christian symbolism, poetry and astrology.

435. Benjamin, Jr., Francis S. "The De quantitatibus stellarum of Thebit ben Kourrah." Essays on Medieval Life and Thought in Honor of Austin Patterson Evans. Edited by John H. Mundy, Richard W. Emery and Benjamin N. Nelson. New York: Columbia University Press, 1955, pp. 91-98.

Describes this work dealing with the sizes of the planets and their spheres, their distances from the earth and the extent of the seven climates. Speculates as to whether or not the work should be attributed to Thābit.

436. Bernard of Verdun. Tractatus super totam astrologiam. Edited with a commentary by Polykarp Hartmann. Werl/Westf.: Dietrich-Coelde, 1961. Pp. 180.

Provides the Latin text of this astronomical treatise by a thirteenth-century Franciscan. Bernard defended the Ptolemaic system and ascribed physical reality to its devices.

437. Birkenmajer, Aleksander. "Robert Grosseteste and Richard Fournival." Études d'histoire des sciences et la philosophie du moyen âge. Edited by Aleksandra Maria Birkenmajer and Jerzy B. Korolec. Studia Copernicana, 1. Warsaw: Polish Academy of Sciences, 1970, pp. 216-221.

Points out that the cosmogony and cosmology in the poem, De vetula, which Fournival ascribed to Ovid, are derived from the "light metaphysics" of Grosseteste. Reprinted from Mediaevalia et Humanistica, 5 (1948), 36-41.

438. Boas, George. "A Fourteenth-Century Cosmology." Proceedings of the American Philosophical Society, 98 (1954), 50-59.

Describes the content of the De macrocosmo by the Venetian, Ludovicus Trevisano, written between 1345-1378. It treats assorted topics such as the nature of number and the continuum, the age and date of creation of the world and issues arising from its possible eternity.

439. Boffito G. and U. Mazzia. "D'un ignoto astronomo del secolo XIV (Pietro di Modena)." Bibliofilia, 8 (1907), 372-383.

An edition of a Theorica planetarum which is actually by Blasius of Parma.

440. Botley, Cicely M. "The Position of Supernova 1006 and the St. Gallen Chronicle." Journal for the History of Astronomy, 7 (1976), 136-140.

Argues that although the best known account of this supernova comes from the Benedictine monastery of St. Gall, the star's extreme southern position raises problems with this report.

441. Campanus of Novara. Campanus of Novara and Medieval Planetary Theory: Theorica planetarum. Edited with an introduction, English translation and commentary by Francis S. Benjamin, Jr. and G.J. Toomer. Madison: University of Wisconsin Press, 1971. Pp. xvi + 490.

The Theorica edited here was the first of the genre to explain Ptolemaic astronomy in numerical detail. It also includes the first European description of an equatorium.

442. Carmody, Francis J. "Leopold of Austria: Li compilacions de la science des estoilles, Books I-III." University of California Publications in Modern Philology 33 (1951), 35-102.

An edition of the first three books of Leopold's treatise dealing respectively with stars and constellations, technical astronomy and a defense of astrology. Originally written in Latin in 1271, it was translated into French before 1324.

443. Carmody, Francis J. "Regiomontanus' Notes on al-Biṭrūjī's Astronomy." Isis, 42 (1951), 121-130.

Gives the Latin text of Regiomontanus' polemic against the homocentric astronomy of al-Biṭrūjī whose work was known in the West through the 1217 translation of Michael Scot.

444. Carmody, Francis J. "The Planetary Theory of Ibn Rushd." Osiris, 10 (1952), 556-586.

Shows that Averroes' views on celestial motion appear in various passages in his commentaries on De caelo and the Metaphysics. Ardent Aristotelian, Averroes violently rejected Ptolemaic astronomy, and had hoped to create an astronomy based on homocentric principles, drawing as did al-Biṭrūjī from the ideas of al-Ṭufayl.

445. Carmody, Francis J., ed. The Astronomical Works of Thābit b. Qurra. Berkeley: University of California Press, 1966. Pp. 262.

Contains editions of Latin versions of works by or attributed to Thābit, chiefly astronomical. Included are the De anno solis, De motu octave spere, De hiis que indigent expositione antequam legatur Almagesti, De recta imaginatione spera, De quantitatibus stellarum, De figura sectore and De imaginibus (on magic images).

446. Curtze, Maximilian, ed. "Der Briefwechsel Regiomontan's mit Giovanni Bianchini, Jacob von Speier und Christian Roder." Urkunden zur Geschichte der Mathematik im Mittelalter und der Renaissance. 2 vols. in 1. Leipzig, 1902. Reprint. New York: Johnson Reprint, 1968, pp. 185-336.

An exchange of letters on astronomical topics.

447. Dales, Richard C. "Medieval Deanimation of the Heavens." Journal of the History of Ideas, 41 (1980), 531-550.

Maintains that while the church fathers tended to reject the ancient notion that celestial bodies were alive or even divine, the notion that the heavens were animated or moved by living beings, intelligences or angels, persisted. However, in the later medieval period, after the crisis of 1277, there were scholastics who accounted for the motion of the heavens in a completely natural and mechanical way.

448. Eastwood, Bruce S. "Notes on the Planetary Configuration in Aberystwyth N.L.W. Ms. 735C, f. 4v." The National Library of Wales Journal, 22 (1981), 129-140.

Describes a page of this MS with planetary configurations and verses containing a peculiar order for the sun, Mercury and Venus. The drawing depends on a tradition stemming from Macrobius and Martianus Capella which featured intersecting circles, but has mistakenly placed the sun between Mercury and Venus.

449. Funkhouser, H. Gray. "A Note on a Tenth-Century Graph." Osiris, 1 (1936), 260-262.

Describes a graph of the tenth or eleventh century in CLM MS 14436 which shows the motion of the planets through the zodiac. See item 1428.

450. Gerard of Cremona. Theorica planetarum Gerardi. Edited by Francis J. Carmody. Berkeley: Printed by the editor, 1942. Pp. 51.

An edition of the Theorica attributed to Gerard, but not by him. See item 484.

451. Goldstein, Bernard R. "On the Theory of Trepidation according to Thābit b. Qurra and al-Zarqāllu and its Implications for Homocentric Planetary Theory." Centaurus, 10 (1965), 232-247.

Describes the model introduced by Thābit to account for the supposed trepidation of the equinoxes. Arzachel's theory differed from that of Thābit and influenced the homocentric astronomy of al-Biṭrūjī.

452. Goldstein, Bernard R. "Some Medieval Reports of Venus and Mercury Transits." Centaurus, 14 (1969), 49-59.

Establishes that there are no well-authenticated reports of transits of Mercury and Venus in the medieval period.

453. Goldstein, Bernard R. "Levi ben Gerson's Analysis of Precession." Journal for the History of Astronomy, 6 (1975), 31-41.

Discusses the opinions of the Hebrew astronomer, Levi ben Gerson, who held that the value of the obliquity of the ecliptic had not varied and who thus rejected the notion of trepidation. In his Astronomy, Levi gave 23;33° as his value for the obliquity and his rate of precession as $1^{o}$ in 67 Egyptian years, 123 days and 16;40 hours, an Egyptian year being exactly 365 days.

454. Goldstein, Bernard R. "Medieval Observations of Solar and Lunar Eclipses." Archives internationales d'histoire des sciences, 29 (1979), 101-155.

Focuses on the chapters dealing with eclipses in the Astronomy of Levi ben Gerson, available to the Latin West in the fourteenth century. Levi discusses 6 lunar and 4 solar eclipses which he observed and compared with theory. Recomputes Levi's data for these based on his procedures and his Tables.

455. Grabmann, Martin. "Die Summa de astris des Gerhard von Feltre, O.P." Mittelalterliches Geistesleben, III. Edited by Ludwig Ott. Munich, 1956. Reprint. Hildesheim: Olms, 1975, pp. 254-279.

Analyzes this astronomical-astrological work by a Dominican schoolman. Its astrological section draws a sharp distinction between natural and superstitious astrology; Gerhard combats the latter. Gives the text of the two prologues to the work and of the section reprobating predictions and chance.

456. Grant, Edward. "Medieval and Seventeenth-Century Conceptions of an Infinite Void Space beyond the Cosmos." Isis, 60 (1969), 39-60.

Suggests that scholastics such as Bradwardine and Oresme who accepted an anti-Aristotelian imaginary infinite void beyond the cosmos were motivated by the implications regarding God's infinite and omnipresent power which was one of the consequences drawn from 1277. God was everywhere, although in a non-dimensional sense. However, the seventeenth century, which accepted a three-dimensional vacuum beyond the earth, ascribed this dimensionality to the deity.

457. Grant, Edward. "Place and Space in Medieval Physical Thought." Motion and Time, Space and Matter: Interrelations in the History of Philosophy and Science. Edited by Peter K. Machamer and Robert G. Turnbull. Columbus: Ohio State University Press, 1976, pp. 137-167.

Argues that the opinion which equated the place of a body with a separate three-dimensional space was continually opposed in scholastic discussions. But emphasis on God's absolute power, an aftermath of the 1277 condemnations, introduced the hypothetical possibility that the deity could produce separate vacua. In the extracosmic realm, theologians such as Bradwardine, Oresme and John de Ripa adduced an infinite void in which God was omnipresent.

458. Hugonnard-Roche, Henri. "Le problème de la rotation de la terre au XIVe siècle." Avant, avec, après Copernic: La representation de l'univers et ses consequénces épistémologiques. Paris: Blanchard, 1975, pp. 61-65.

Contrasts the views on the possible rotation of the earth by two fourteenth-century schoolmen, Nicole Oresme and Themon Judaeus.

459. Klug, Rudolf. Johannes von Gmunden, der Begründer der Himmelskunde auf deutschen Boden. Sitzungsberichte der Akademie der Wissenschaften in Wien, phil.-hist. Klasse, 222. Vienna: Hölder-Pichler-Tempsky, 1943. Pp. 93.

Surveys the life, career and works of this prominent Viennese arts master and astronomer with particular attention to his astronomical tables, instrument treatises and calendars.

460. Kren, Claudia. "Homocentric Astronomy in the Latin West: The De reprobatione ecentricorum et epiciclorum of Henry of Hesse." Isis, 59 (1968), 269-281.

Discusses the futile effort of the German scholastic Henry of Hesse (Langenstein) to substitute astronomical models of his own for those of Ptolemy. Henry's "system" is a combination of the Eudoxian spheres and the trepidation model of Thābit ibn Qurra.

461. Kren, Claudia. "A Medieval Objection to 'Ptolemy'." British Journal for the History of Science, 4 (1969), 378-393.

Analyzes some of the criticisms of purported Ptolemaic theory in the De reprobatione ecentricorum et epiciclorum of the German schoolman, Henry of Langenstein. Henry seems to have found the astronomical theory to which he objects in the Theorica planetarum literature, especially the Theorica attributed to Gerard of Cremona.

462. Kren, Claudia. "The Rolling Device of Naṣīr al-Dīn al-Tūsī in the De sphera of Nicole Oresme?" Isis, 62 (1971), 490-498.

Suggests, based on a passage in Oresme's Questions on the Sphere of Sacrobosco, that some version of the device invented by the eastern astronomer, al-Tūsī, may have been known to Oresme.

463. Kren, Claudia. "Planetary Latitudes, the Theorica Gerardi, and Regiomontanus." Isis, 68 (1977), 194-205.

Analyzes the passage in Regiomontanus' fulmination against the section of the Theorica attributed to Gerard which treats planetary latitudes. "Gerard's" latitude theory, as was often the case in medieval treatments of the topic, was a confusion of Ptolemaic and Indian notions. See item 490.

464. Lai, Tyrone. "Nicholas of Cusa and the Finite Universe." Journal of the History of Philosophy, 11 (1973), 161-167.

Concludes from the interpretation of "infinite" in Cusa's work, that the concept of an infinite universe in Cusa's De docta ignorantia implies that the universe is finite in extent, but is infinite in the privative sense that number is infinite; the universe is thus infinite in that it could be made greater than it is by God's power.

465. Litt, Thomas. Les corps célestes dans l'univers de Saint Thomas d'Aquin. Paris: Béatrice-Nauwelaerts, 1963. Pp. 408.

Argues that, metaphysically speaking, Aquinas had a coherent conception of the universe which from the supreme cause downward comprised a chain of causes including the celestial spheres which exercised an influence on the inferior world. In the area of astronomy, maintains that Aquinas, while aware of the hypothetical nature of astronomical theories, remained indifferent to the dilemma presented by the Eudoxian-Aristotelian system vis-à-vis that of Ptolemy.

466. Lorch, Richard P. "The Astronomy of Jābir ibn Aflaḥ." Centaurus, 19 (1975), 85-107.

Jābir wrote an influential critique of the Almagest in nine books which was translated by Gerard of Cremona in 1175. The work contains trigonometric methods not in the Almagest; its most famous criticism of Ptolemy is Jābir's rejection of the Almagest's location of Mercury and Venus below the sun.

467. Markowski, Mieczysław. "Die kosmologischen Anschauungen des Prosdocimo de' Beldomandi." Studi sul XIV secolo in memoria Anneliese Maier. Edited by Alfonso Maierù and Agostino Paravicini Bagliani. Rome: Edizioni di storia e letteratura, 1981, pp. 265-273.

Discusses the cosmological notions in Prosdocimo's commentary on the Sphere of Sacrobosco. Although Prosdocimo followed the commented text, he posed a number of alternative ideas not part of the usual cosmological traditions.

468. Michel, Henri. "Sur l'origine de la théorie de la trépidation." Ciel et terre, 66 (1950), 227-234.

Discusses ancient theories of the oscillation of the eighth sphere, and the theory associated with Thābit ibn Qurra. Argues that the difficulties in Thābit's work may be due to the translator and that possibly the treatise is not by Thābit. Finds the origin of the theory of trepidation in observations of the ortive amplitude (and its western equivalent), in a confusion of coordinate systems as well as in the adoption by Thābit of a notion Ptolemy had invented to account for the oscillation of epicycles in his latitude theory.

469. Millás Vallicrosa, J.M. Estudios sobre Azarquiel. Madrid: Consejo superior de investigaciones científicas, 1943-1950. Pp. xii + 530.

Treats the works of the eleventh-century Spanish-Arabic astronomer, Arzachel, Azarquiel (al-Zarqālī) of Cordoba best known in the Christian West as the inventor of the saphea and for the Toledan Tables which are ascribed to him. Analyzes the works of Arzachel, their influence on Arabic authors and their use by Christian composers of tables and astronomical treatises. See item 212.

470. Moesgaard, Kristian P. "Thābit ibn Qurra between Ptolemy and Copernicus." Avant, avec, après Copernic: La representation de l'univers et ses conséquences épistémologiques. Paris: Blanchard, 1975, pp. 67-70.

Points out that Arabic astronomers who made new observations had to explain changes in parameters Ptolemy had assumed constant such as the length of the tropical year. An effort to do this was made by Thābit in his De anno solis.

471. Mundy, John. "John of Gmunden." Isis, 34 (1942-43), 196-205.

Discusses John's life and astronomical works, especially his tables and almanachs.

472. Nallino, C.A. "Il Gherardo Cremonese autore della Theorica planetarum deve ritenersi essere Gherardo Cremonese da Sabbioneta." Rendiconti, Real Accademia dei Lincei. Classe scienze morali, storiche e filologiche, ser. 6, 8 (1932), 386-404.

Responds to arguments which Duhem adduced in attributing this Theorica to the twelfth-century translator, Gerard of Cremona. The author believed the work was due to the thirteenth-century Gerard of Sabbioneta. Reprinted in Raccolta di scritti editi e inediti, 6. Rome, 1948.

473. Neugebauer, O. "Thābit ben Qurra 'On the Solar Year' and 'On the Motion of the Eighth Sphere'." Proceedings of the American Philosophical Society, 106 (1962), 264-299.

Translations with commentary of these works based on the Latin text of F.J. Carmody. See item 445.

474. Neugebauer, O. and O. Schmidt. "Hindu Astronomy at Newminster in 1428." Annals of Science, 8 (1952), 221-228.

Points out that the methods employed in this treatise of astronomical computations for 1428 at Newminster are related to those of the Hindu astronomical work, the Sūrya Siddhānta, as shown, for example, in its use of the Hindu astronomical era, the times of ascension of arcs of the ecliptic and the use of the Hindu norm, R=150. This information was transmitted through Arabic sources. See item 506.

475. Newton, Robert R. Medieval Chronicles and the Rotation of the Earth. Baltimore: Johns Hopkins Press, 1972. Pp. xvii + 825.

Contains a large number of accounts of solar eclipses from medieval records.

476. North, John D. "Walter of Odington and the History of the Eighth Sphere." Actes du XIIe congrès international d'histoire des sciences, IIIA. Paris: Blanchard, 1971, pp. 113-118.

Argues that Walter of Odington, monk of Eversham, in his observations and his star catalogue showed some indication of originality.

477. North, John D. "Kinematics--More Ethereal than Elementary." Machaut's World: Science and Art in the Fourteenth Century. Edited by Madeleine Pelner Cosman and Bruce Chandler. New York: New York Academy of Sciences, 1978, pp. 89-102.

Argues that Aristotle's division of his universe into two regions is reflected in the division between schoolmen who wrote on Ptolemaic astronomy and their counterparts who devoted themselves to terrestrial kinematics. However, there were perceptible interactions between the two applications of kinematics, and those concerned with the kinematics of the heavens were as inventive as were the natural philosophers and mathematicians.

478. dall'Olmo, Umberto. "Meteors, Meteor Showers and Meteorites in the Middle Ages: From European Medieval Sources." Journal for the History of Astronomy, 9 (1978), 123-134.

Lists accounts of meteors and associated phenomena compiled from European annals and chronicles from the fifth to the fifteenth centuries with brief summaries of the records.

479. dall'Olmo, Umberto. "Latin Terminology Relating to Aurorae, Comets, Meteors and Novae." Journal for the History of Astronomy, 11 (1980), 10-27.

Provides a glossary of astronomical terms drawn from medieval and classical sources relating to the four phenomena given above.

480. Oresme, Nicole. Nicole Oresme and the Kinematics of Circular Motion: Tractatus de commensurabilitate vel incommensurabilitate motuum celi. Edited with an introduction, English translation and commentary by Edward Grant. Madison: University of Wisconsin Press, 1971. Pp. xx + 415.

The treatise is concerned with the consequences arising from the circular motion of bodies moving with velocities which are either commensurable or incommensurable. It would appear that, given these alternatives, Oresme preferred to opt for the incommensurability of celestial motions thus destroying the predictive claims of astrologers.

481. Pedersen, Olaf. "The Theorica planetarum Literature of the Middle Ages." Classica et Mediaevalia, 23, (1962), 225-232.

Discusses several of the astronomical treatises with this title, but especially the short extremely popular Theorica used in university teaching along with Sacrobosco's Sphere, often found with the latter in the MSS and sometimes connected with Gerard of Cremona. Although the work's influence declined in the second half of the fifteenth century, it continued to appear in printed versions.

482. Pedersen, Olaf. "The Corpus astronomicum and the Traditions of Mediaeval Latin Astronomy." Colloquia Copernicana, 3. Warsaw: Polish Academy of Sciences, 1975, pp. 57-97.

Sees the decisive point in European astronomy in the twelfth-century return of Greek astronomy and the advent of Arabic materials, furthered by the rise of the university. Argues that astronomical teaching and interests in the university milieu can best be evaluated by an examination of representative astronomical codices containing required manuals bound with other astronomical treatises and written in the same hand or hands.

483. Pedersen, Olaf. "The Decline and Fall of the Theorica planetarum." Science and History: Studies in Honor of Edward Rosen. Edited by Erna Hilfstein, Pawel Czartoryski and Frank D. Grande. Studia Copernicana, XVI. Warsaw: Ossolineum, 1978, pp. 157-185.

Traces the late medieval fortunes of this ubiquitous Theorica. The advent of printing in the fifteenth century served to keep the work alive; however, at this time astronomical activity at the University of Vienna, especially the new Theorica by Peurbach, provided a more competent rival. Regiomontanus, at his press at Nuremberg, both popularized Peurbach's work and, in his polemic against the older Theorica, furthered its decline.

484. Pedersen, Olaf. "The Origins of the Theorica planetarum." Journal for the History of Astronomy, 12 (1981), 112-123.

Argues from a survey of the medieval astronomical literature and from manuscript evidence that this extremely popular Theorica planetarum, long used as a text in university teaching, was not, despite some medieval and modern attributions, written by Gerard of Cremona, but by an unknown teacher of astronomy writing about the mid-thirteenth century.

485. Pereira, Michela. "Campano da Novara, autore dell'Almagestum parvum." Studi medievali, 19 (1979), 769-776.

Argues, based on MS attribution, sources used and the character of the work, that the Almagestum parvum ascribed to several authors, but frequently to Geber (Jābir ibn Aflaḥ), is an early work by Campanus, preliminary to his Theorica.

486. Porter, N.A. "The Nova of A.D. 1006 in European and Arab Records." Journal for the History of Astronomy, 5 (1974), 99-104.

Attests that this nova is well-documented in European, Islamic and far eastern records, although no observation from northern Europe has been noted. Despite its reported brightness, no reasonable candidate for the remnant of the nova as yet has been discovered.

487. Poulle, Emmanuel. Astronomie théorique et astronomie pratique au moyen âge. Paris: Palais de la Découverte, 1967. Pp. 32.

Points out that medieval theoretical and practical astronomy were not opposed, but complementary, and that neither was observational in nature. The goal of the astronomer was to determine the exact positions of celestial bodies at a given time; this was done by two general methods--astronomical tables and instruments (equatoria). Tables involved the possibility of numerous errors of calculation and required considerable time. Equatoria made it possible to dispense with tables, for the most part, and although the results were approximate, errors were avoided.

488. Poulle, Emmanuel. "La théorie épicyclique selon Ptolémée au moyen âge." Avant, avec, après Copernic: La representation de l'univers et ses conséquences épistémologiques. Paris: Blanchard, 1975, pp. 51-60.

Confirms that after the twelfth-century return of Greek technical astronomy, the epicycles and other devices of Ptolemy, with minor additions concerning precession, maintained their supremacy for centuries. Provides a brief description of Ptolemaic models.

489. Price, Betsy Barker. "The Physical Astronomy and Astrology of Albertus Magnus." Albertus Magnus and the Sciences: Commemorative Essays. Edited by James A. Weisheipl. Toronto: Pontifical Institute of Mediaeval Studies, 1980, pp. 156-185.

Argues that Albert distinguished between astronomy and astrology. He accepted the notion of celestial influences, but rejected their effect on free will. Astronomical passages in his commentaries on the libri naturales and the Metaphysics show his familiarity with the Eudoxian-Aristotelian system, Ptolemy and the model of al-Biṭrūjī. While recognizing the hypothetical nature of astronomical systems, he had a penchant for mechanical models and physical theories; technical astronomy played little part in his astronomical descriptions.

490. Regiomontanus. Johannis Regiomontani contra Cremonesia deliramenta in theorica planetarum disputatio. Milliaria, 10. Aalen: Zeller, 1972.

A facsimile reprint of Regiomontanus' treatise against the Theorica associated with Gerard of Cremona, inspired by his distress at the low state of contemporary university instruction in astronomy.

491. Roskińska, Grażyna. "Sandivogius de Czechel et l'école astronomique de Cracovie vers 1430." Organon, 9 (1973), 217-229.

Points out how the commentary on the Theorica planetarum associated with Gerard of Cremona by this arts master who taught at Cracow in 1430, and perhaps in 1431, provides a vivid picture of the sources available to and the astronomical interests of the Cracow school.

492. Rosińska, Grażyna. "Nasīr al-Dīn al-Ṭūsī and Ibn al-Shāṭir in Cracow?" Isis, 75 (1974), 239-243.

Argues that the rolling device of al-Ṭūsī and the lunar model of Ibn al-Shāṭir (used by Copernicus who had studied at Cracow) may be reflected in the use of two epicycles in the treatment of the moon found in the commentaries of two fifteenth-century Cracow masters: Sandivogius (on the Gerard Theorica planetarum) and Adalbertus of Brudzewo (on Peurbach's Theorice nove planetarum).

493. Silverstein, Theodore. "Daniel of Morley, English Cosmogonist and Student of Arabic Science." Mediaeval Studies, 10 (1948), 179-196.

Affirms that while the cosmogony of the Liber de naturis inferiorum et superiorum is dependent on the Timaeus and Chartrian authors, Daniel, having studied with Gerard of Cremona at Toledo, was familiar with Arabic works. Argues that a fusion between Arabic and Chartrian cosmological ideas may have existed at Toledo.

494. Stahlman, William D. "Astronomical Dating Applied to a Type of Astrological Illustration." *Isis*, 47 (1956), 154-160.

Provides examples of how astronomical data, such as the location of celestial bodies in the zodiac found in astrological drawings, can be used for dating.

495. Stock, Brian. "Hugh of St. Victor, Bernard Silvester and MS Trinity College, Cambridge O. 7. 7." *Mediaeval Studies*, 34 (1972), 152-173.

Identifies the little cosmological treatise in the MS as a paraphrase of the *De tribus diebus* attributed to Hugh of St. Victor. The scientific content taken from Hugh is emphasized at the expense of the theological, and natural processes are stressed in the manner of William of Conches, Thierry of Chartres and Bernard Silvestris, especially the latter.

496. Sudhoff, Karl. "Daniel von Morley *Liber de naturis inferiorum et superiorum*, nach der Handschrift Cod. Arundel 377 des Britischen Museums zum Abdruck gebracht." *Archiv für Geschichte der Naturwissenschaften und der Technik*, 8 (1918), 1-40.

Latin text of the treatise.

497. Swerdlow, Noel. "Al-Battānī's Determination of the Solar Distance." *Centaurus*, 17 (1973), 97-105.

Discusses al-Battānī's redetermination of the solar distance (Chapter 30 of the *Opus astronomicum*); his result of 1146 terrestrial radii was as well-known as Ptolemy's 1210 tr from *Almagest* V, 15. See item 428.

498. Talbot, Charles H. "Simon Bredon (c. 1300-1372), Physician, Mathematician and Astronomer." *British Journal for the History of Science*, 1 (1962-63), 19-30.

Discusses Bredon's career and works. Associated with Merton College, Bredon's early scientific concerns were medical, but his later career was apparently devoted to mathematics and astronomy. Among his extant treatises are the *Trifolium* (on medicine), commentaries on the *Arithmetic* of Boethius and on the *Almagest*, as well as minor astronomical works.

499. Themon Judaeus. L'oeuvre astronomique de Themon Juif, maître parisien du XIVe siècle. Edited with a commentary by Henri Hugonnard-Roche. Geneva: Droz, 1973. Pp. 429.

Analyzes Themon's commentary or paraphrase and questions on the De spera of Sacrobosco, his questions on the Meteorology and his question on the moon's motion disputed at Erfurt in 1350. Provides an edition of the text of this last treatise.

500. Thomson, S. Harrison. "The Text of Grosseteste's De cometis." Isis, 19 (1933), 19-25.

Presents the text of this work in a more complete form than that in Baur's edition. See items 188, 264.

501. Thorndike, Lynn. "Peter of Limoges on the Comet of 1299." Isis, 36 (1945), 3-7.

Gives the Latin text and English translation of a comet treatise by Peter of Limoges, canon of Evreux. See item 505.

502. Thorndike, Lynn. "Astronomical Observations at Paris from 1312 to 1315." Isis, 38 (1947-48), 200-205.

Points out that observations of positions of the planets and fixed stars were made between these dates for the meridian of Paris. Gives the Latin text of the relevant folios of Oxford, Corpus Christi MS 144.

503. Thorndike, Lynn. "Thomas Werkwoth on the Motion of the Eighth Sphere." Isis, 39 (1948), 212-215.

Discusses material on the eighth sphere written near the end of 1395 in which Werkwoth describes the theories of precession known to him.

504. Thorndike, Lynn. The Sphere of Sacrobosco and its Commentators. Chicago: University of Chicago Press, 1949. Pp. x + 496.

Latin text and English translation with introductory commentary of the Tractatus de spera. Includes the text with translation of the commentary by Robertus Anglicus and the texts of commentaries by Michael Scot,

Cecco d'Ascoli and an anonymous author. Appendixes provide excerpts from the _De spera_ of Pecham and from other anonymous commentaries.

505. Thorndike, Lynn, ed. _Latin Treatises on Comets between 1238 and 1368 A.D._ Chicago: University of Chicago Press, 1950. Pp. 275.

Presents the texts of several comet treatises in chronological order and includes translations of the comet passages from the commentaries on the _Meteorology_ by Albertus Magnus and Aquinas.

506. Thorndike, Lynn. "Astronomical and Chronological Calculations at Newminster in 1428." _Annals of Science_, 7 (1951), 275-283.

Provides the text and English translation of this work in Bodleian, Ashmole MS 191.II. See item 474.

507. Thorndike, Lynn. "Prediction of Eclipses in the Fourteenth Century." _Isis_, 42 (1951), 301-302.

Reports a list of predictions of eight solar eclipses between 1366 and 1386 in a MS at Utrecht.

508. Thorndike, Lynn. "Eclipses in the Fourteenth and Fifteenth Centuries." _Isis_, 48 (1957), 51-57.

Describes the eclipse treatises in 5 Sloane MSS in the British Library for the period 1327-1462 by Walter of Elevedene, John Somer and Nicholas of Lynn.

509. Thorndike, Lynn. "Johannes de Luna Theutonicus, about A.D. 1305." _Isis_, 56 (1965), 207-208.

Reports briefly on the astronomical observations made by a Johannes de Luna in a MS in the National Library at Vienna. He may be identical with a "Joannes a Luna" who taught at Bologna.

510. Vennebusch, Joachim. "Bemerkung zum _Tractatus de stellis_ des Robert Holkot." _Bulletin de philosophie médiévale_, 20 (1978), 75.

Argues that the Holkot treatise was actually intended to comprise the last question of his commentary on Book II of the _Sentences_.

511. Weisheipl, James A. "The Celestial Movers in Medieval Physics." The Dignity of Science. Edited by James A. Weisheipl. Washington, D.C.: Thomist Press, 1961, pp. 150-190.

Compares the views of three Dominican masters, Albertus Magnus, Kilwardby and Thomas Aquinas on the nature of celestial movers--whether separate intelligences, angels or an innate tendency to move circularly --arising from a list of questions placed by the Master General of their order in 1271.

512. Zalcman, Lawrence. "The Great Schism and the Supernova of 1054." Physis, 21 (1979), 55-59.

Suggests that mention of the supernova of 1054 could have been suppressed because it may have been interpreted as a harbinger of disaster. The supernova occurred about the same time in 1054 as the schism between the two branches of the Catholic church.

513. Zinner, Ernst. "Magister Alard von Diest und die Pariser Beobachtungen von 1312-1315." Isis, 42 (1951), 38-40.

Suggests that the 1307-08 star catalogue of 14 stars of Alard of Diest in Brabant was used by Parisian observers in the years 1312-15.

514. Zinner, Ernst. Leben und Wirken des Joh. Müller von Königsberg genannt Regiomontanus. 2d rev. ed. Osnabrück: Zeller, 1968. Pp. 402.

An intellectual biography which discusses the astronomical activities and works of Regiomontanus in the context of contemporary astronomical and mathematical developments, university teaching, and book publishing. Provides information concerning other Viennese astronomers such as John of Gmunden and Peurbach.

## ASTRONOMICAL TABLES

515. Harper, Richard I. "Prophatius Judaeus and the Medieval Astronomical Tables." Isis, 62 (1971), 61-68.

Compares the ephemerides in the Almanach perpetuum with the values in the Toledan, Alphonsine and Toulousan Tables so as to ascertain which set was the most likely source for Prophatius. See item 534.

516. Harper, Richard I. "The Astronomical Tables of William Rede." Isis, 66 (1975), 369-378.

Outlines the contents of the canons accompanying Rede's adaptation of the Alphonsine Tables for the meridian of Oxford. A sample comparison of Rede's tables with the Parisian adaptation of the Alphonsine Tables made by John de Lineriis reveals close agreement.

517. Kennedy, E.S. "A Survey of Islamic Astronomical Tables." Transactions of the American Philosophical Society, n.s., vol. 46, part 2. Philadelphia: American Philosophical Society, 1956, pp. 121-177.

Surveys the contents of a number of Islamic zijes (astronomical tables); two of these, those of al-Khwārizmī and al-Battānī, were known in the Latin West.

518. Kennedy, E.S. and Mardiros Janjanian. "The Crescent Visibility Table in al-Khwārizmī's zij." Centaurus, 11 (1965), 73-78.

Discusses a lunar visibility table computed on the basis of Indian visibility theory to be used at the latitude of northern Spain.

519. Kennedy, E.S. and Walid Ukashah. "al-Khwārizmī's Planetary Latitude Tables." Centaurus, 14 (1969), 86-96.

Describes and analyzes the latitude tables in the astronomical tables of al-Khwārizmī and the Indian latitude model on which they are based.

520. al-Khwārizmī. Die astronomischen Tafeln des Muḥammad ibn Mūsā al-Khwārizmī in der Bearbeitung des Maslama ibn Aḥmed al-Madjrīṭī und der lateinischen Übersetzung des Athelard von Bath. Edited by Heinrich Suter based on preliminary work of A. Björnbo and R. Besthorn. Copenhagen: Høst & Søn, 1914. Pp. xxv + 237. See item 521.

521. al-Khwārizmī. The Astronomical Tables of al-Khwārizmī. Translated with commentaries of the Latin version edited by H. Suter supplemented by Corpus Christi College MS 283 by Otto Neugebauer. Kongelige Danske Videnskabernes Selskab. Historisk-filosofiske Skrifter, 4, 2. Copenhagen: Munksgaard, 1962. Pp. 247.

English translation with commentary of the astronomical tables of al-Khwārizmī in the version of Maslama al-Majrīṭī, translated into Latin by Adelard of Bath, which Suter published based on preliminary work by Björnbo and Besthorn, with material from an incomplete version identified too late to have been used by Suter.

522. Kunitzsch, Paul. Typen von Sternverzeichnissen in astronomischen Handschriften des zehnten bis vierzehnten Jahrhunderts. Wiesbaden: Harrasowitz, 1966. Pp. 137.

Presents star catalogues from the MSS, classified into 17 types by various factors such as the coordinates employed, the names of the stars, the period and title of the list, etc. Those star lists based on the Almagest and the Alphonsine Tables have been excluded as have a few other short lists. Catalogues from Arabic sources are included when they provided the basis for Latin star tables.

523. Millás Vallicrosa, J.M. El libro de los fundamentos de las tablas astronómicas de R. Abraham ibn 'Ezra. Madrid: Consejo superior de investigaciones científicas, 1947. Pp. 171.

An edition of the text of astronomical canons. Discusses sources of the text, its place within the scientific ambience of the time and its relation with other works of Ibn 'Ezra. The canons mention the discrepancies between the astronomical systems known to the twelfth century and attempt to harmonize them. See item 212.

524. Millás Vallicrosa, J.M. Las tablas astronómicas del rey Don Pedro el Ceremonioso. Madrid: Consejo superior de investigaciones científicas, 1962. Pp. 240.

Provides the texts of the canons to the astronomical tables of Peter of Aragon in the Hebrew and Catalan version and the text of the tables themselves based on these versions. Gives also the Latin text of the canons in the redaction of Jacob Corsino. The Latin version of the tables has not been found. See items 213, 259.

525. Millás Vallicrosa, J.M. "Ibn al-Muṭannā' et le prologue de son commentaire aux tables d'al-Jwārizmī." Mélanges Alexandre Koyré, I. Paris: Hermann, 1964, pp. 387-396.

Maintains that the prologue to this commentary in the Latin version of Hugo Sanctellensis reveals it is not by al-Fargānī. Hugo's prologue utilizes material from both the text of ibn-Muṭannā' and from the latter's own prologue. See item 526.

526. Millás Vendrell, Eduardo. El comentario de Ibn al-Muṭannā' a las Tablas Astronómicas de al-Jwārizmī. Critical edition of the Latin text in the version of Hugo Sanctallensis. Madrid: Consejo superior de investigaciones científicas, 1963. Pp. 209.

Argues that the commentary of Ibn al-Muṭannā' was based on the major recension of al-Khwārizmī's tables unlike the version of Maslama of Cordoba which was translated by Adelard of Bath.

527. North, John D. "The Alfonsine Tables in England." Prismata: Naturwissenschaftsgeschichtliche Studien. Festschrift für Willy Hartner. Edited by Y. Maeyama and W.G. Salter. Wiesbaden: Steiner, 1977, pp. 269-301.

Argues that the Alfonsine Tables, received at Paris in the early fourteenth century, were reshaped and modified by astronomers such as John de Muris and John de Lineriis; when transferred to England by 1330, they were further modified as the Oxford Tables of 1348 and the work of the Mertonian astronomer John Killingworth attest.

528. Poulle, Emmanuel. "Astrologie et tables astronomiques au XIIIe siècle: Robert Le Febvre et les tables de Malines." Bulletin philologique et historique du comité des travaux historiques et scientifiques, (1964), 793-831.

Treats the use of the Malines Tables, constructed by Henry Bate for the meridian of his native city, by Bate himself, but particularly by Le Febvre in the preparation of both his natal horoscope and that constructed at the beginning of his thirtieth year. A comparison of values of the regular motions of the planets given by Le Febvre for his birth date with those calculated from two versions of the Malines Tables reveal discrepancies due to misreadings of a defective MS version of the tables. Calculation of true positions at Le Febvre's birth according to the mean motions and mean arguments he determined shows various types of errors which provided ready pitfalls for the medieval astronomer.

529. Poulle, Emmanuel. A propos des tables astronomiques de Pierre d'Aragon. Coimbra: Junta de investigações. do ultramar, 1966. Pp. 15.

Discusses the peculiarities found in the tables of Peter of Aragon of 1320. There were three versions of these tables in Catalan, Hebrew and Latin; the Latin version has not been found. Points out the errors and anomalies in the mean motion and mean argument tables as well as the unusual format of the tables of equations for finding true place. These tables appear in part in a MS written for or by a fifteenth-century doctor, Jean Thibaut, although not in a form transmitted by the Catalan or Hebrew versions. See item 525.

530. Poulle, Emmanuel. "Jean de Murs et les tables Alphonsines." Archives d'histoire doctrinale et littéraire du moyen âge, 47 (1980), 241-271.

John's treatise, Expositio intensionis regis Alfonsii circa tabulas ejus was written in 1321. This marks the date of the first certain appearance of the Alphonsine Tables at Paris. Provides the text and a running commentary in the footnotes.

531. Poulle, Emmanuel and Owen Gingerich. "Les positions des planètes au moyen âge: Application du calcul électronique aux tables Alphonsines." Comptes rendus, Academie des Inscriptions et Belles-lettres, (1967), 531-548.

Points out the value of computer projects since the early 1960's, especially the tables of planetary positions provided by Bryant Tuckerman, to the medieval historian in reconciling problems connecting given astronomical information and exact chronology or vice versa. Describes a project which applied computer techniques to an analysis of the *Alfonsine Tables* based on medieval methods not modern formulas. This analysis should be of use in determining the diffusion of the *Alfonsine Tables* and the relationships between various medieval tables based on them or on the older *Toledo Tables*.

532. Rosen, Edward. "The Alfonsine Tables and Copernicus." *Science, Medicine and the University: 1200-1500. Essays in Honor of Pearl Kibre*, II. Edited by Nancy G. Siraisi and Luke Demaitre. *Manuscripta*, 20 (1976), pp. 163-174.

Points out that these thirteenth-century astronomical tables, translated from Arabic into Castillian under the sponsorship of Alfonso X, surfaced at Paris and Oxford in the first half of the fourteenth century. They were subsequently printed, but although Copernicus owned a copy of them, he used them neither in the *Commentariolus* nor the *De revolutionibus*.

533. Toomer, G.J. "A Survey of the *Toledan Tables*." *Osiris*, 15 (1968), 5-174.

Describes each of the tables in the *Toledan Tables* and reproduces some in full. Concludes that there is little that is independent in the *Toledan Tables*; most were copied from al-Khwārizmī and al-Battānī. See item 535.

534. Toomer, G.J. "Prophatius Judaeus and the *Toledan Tables*." *Isis*, 64 (1973), 351-355.

Presents computational evidence that the *Almanach* of Prophatius was based on the *Toledan Tables*.

535. Zinner, Ernst. "Die Tafeln von Toledo (*Tabulae Toletanae*)." *Osiris*, 1 (1936), 747-774.

Lists a number of MSS of Latin translations of the *Toledo Tables*, as well as the titles of the tables and a brief explanation of their content, taken from a version in the National Library at Vienna. Argues that the tables were constructed in Toledo by Arabic scholars, based on those of al-Khwārizmī and al-Battānī. They were translated and adapted for use in the Christian West during the twelfth century.

## ASTRONOMICAL INSTRUMENTS, SUNDIALS AND CLOCKWORK

536. Bedini, Silvio A. and Francis R. Maddison. Mechanical Universe: The Astrarium of Giovanni de' Dondi. Transactions of the American Philosophical Society, n.s., 56, part 5. Philadelphia: American Philosophical Society, 1966. Pp. 69.

Traces the antecedents and fortunes of the weight-driven, complex astronomical clock designed and built by de'Dondi between 1348 and 1364, and described in his Tractatus astrarii. The mechanism was highly regarded during the fourteenth to early sixteenth centuries, although it dropped out of sight by about 1530. Renewed interest in the present century has led to two reconstructions.

537. Benjamin, Jr., Francis S. "John of Gmunden and Campanus of Novara." Osiris, 11 (1954), 221-246.

Provides the text of John's equatorium treatise, taken with only minor changes from Campanus' Theorica. See item 441.

538. Bergmann, Werner. "Der Traktat De mensura astrolabii des Hermann von Reichenau." Francia, 8 (1980), 65-103.

Presents new information as to the Latin sources of Hermann's treatise drawn from MS traditions concerning the star table in the work. Concludes that it is not possible to find the connection between Arabic treatises on the astrolabe of the tenth century and the first Latin translations, but that the oldest Latin astrolabe treatises are of French-Lotharingian origin.

539. Destombes, Marcel. "Un astrolabe carolingien et l'origine de nos chiffres arabes." Archives internationales d'histoire des sciences, 15 (1962), 3-45.

Announces the discovery of an astrolabe constructed to be used at Barcelona, at Vich or Ripoll, between the mid-tenth century and 986. It falls in the milieu of astronomical activity at that time in the Spanish March (Barcelona) which transmitted Arabic information on the astrolabe beyond the Pyrenees. Epigraphical evidence, taken from the astrolabe and independent documentation, suggests an hypothesis as to the origin of Arabic numerals.

540. Destombes, M. "La diffusion des instruments scientifiques du haut moyen âge au XVe siècle." Cahiers histoire mondiale, 10 (1966), 2-51.

Traces the development of medieval astronomical instruments of all kinds which reached its highest peak in Arabic cultural areas and was, in part, transmitted to the Latin West where, during the fourteenth and fifteenth centuries, astronomers constructed not only instruments based on Arabic inspiration, but created new types.

541. Drecker, J. "Hermannus Contractus über das Astrolabe." Isis, 16 (1931), 200-219.

Maintains that the astrolabe was transmitted to the West from Arabic Spain by the eleventh century among translations of Arabic astronomical works in Latin. One of those who made use of these early materials was Hermannus Contractus of Reichenau. Gives the text of Hermannus' De mensura astrolabii.

542. Goldstein, Bernard R. "Levi ben Gerson: On Instrumental Errors and the Transverse Scale." Journal for the History of Astronomy, 8 (1977), 102-112.

Points out that while Levi used a transversal scale for linear measures in his description of the Jacob Staff, he also invented a transversal scale for subdividing angles to deal with errors arising in the use of astrolabes.

543. Gunther, R.T. Astrolabes of the World. 2 vols. Oxford: Oxford University Press, 1932.

Vol. II discusses European astrolabes in the Lewis Evans collection at Oxford and in other collections.

544. Hahn, Nan L. Medieval Mensuration: Quadrans vetus and Geometrie due sunt partes principales.... Transactions of the American Philosophical Society, 72, Part 8. Philadelphia: American Philosophical Society, 1982. Pp. lxxxv + 204.

Presents the text of a thirteenth-century treatise on the "old quadrant" of Arabic provenance, written at Montpellier likely 1250-1280, with an analysis of the treatise and its sources. Believes it should be attributed to Johannes Anglicus, not Robert. Finds that the bulk of the material in Quadrans vetus stems from two sources: a twelfth-century Practica geometria, author unknown, with an almost identical incipit: "Geometrie due sunt partes principales...." also edited here. The second source is an old quadrant work in the Libros del saber.... See items 429, 553, 575 and 578.

545. Hartmann, J. "Die astronomischen Instrumente des Kardinals Nikolaus Cusanus." Abhandlungen der Königlichen Gesellschaft der Wissenschaften zu Göttingen, math.-phys. Klasse, n.f., 10, 6 (1919), 1-56.

Describes Cusa's astronomical instruments, conserved at his birthplace at Cues. These include a torquetum, a large wooden celestial globe and an astrolabe, all purchased by Cusa at Nuremberg, as well as a small copper celestial globe.

* Jordanus de Nemore. Jordanus de Nemore and the Mathematics of Astrolabes: De plana spera. Edited with introduction, translation and commentary by Ron B. Thomson. Toronto: Pontifical Institute of Mediaeval Studies, 1978. Pp. x + 238.

Cited as item 870.

546. Kren, Claudia. "The Traveler's Dial in the Late Middle Ages: The Chilinder." Technology and Culture, 18 (1977), 419-435.

Discusses the origin (Arabic) and reception in the West of this portable sundial independent of latitude, and describes its construction and use as seen in the treatise on the instrument by John of Gmunden.

547. Kunitzsch, Paul. "On the Authenticity of the Treatise on the Composition and Use of the Astrolabe Ascribed to Messahalla." Archives internationales d'histoire des sciences, 31 (1981), 42-62.

Draws attention to the problem of whether this astrolabe treatise can actually be ascribed to Messahalla, an attribution which has been questioned by several scholars. Concludes that the treatise is not by Messahalla nor is the work on the composition and use of the astrolabe by Rudolf of Bruges a translation from Arabic; it is rather a work by Rudolf himself. For doubts as to the authenticity of the Messahalla treatise, see items 211, p. 154; 212, p. 106f; 213, p. 72f; 522, p. 51n1 and 560, p. 85.

548. Lorch, Richard P. "The Astronomical Instruments of Jābir ibn Aflaḥ and the Torquetum." Centaurus, 20 (1976), 11-34.

Points out that in his work on the Almagest, Jābir described an instrument similar to the European torquetum, but that in the Latin version of Jābir's work (translated by Gerard of Cremona), an instrument not unlike the triquetum of Almagest V, 12 is described.

549. Lorch, Richard P. "The Sphera solida and Related Instruments." Centaurus, 24 (1980), 153-161.

Describes the solid sphere as it appears in the Latin western tradition. Its use was similar to that of the plane astrolabe. It represents one of a class of three-dimensional astronomical instruments such as the spherical astrolabe or the armillary of the Almagest. These instruments were well-known in the Islamic world where new types were invented.

550. Maddison, Francis. "Early Astronomical and Mathematical Instruments: A Brief Survey of Sources and Modern Studies." History of Science, 2 (1963), 17-50.

Reviews the work of modern scholarship on early scientific instruments and discusses resources such as instrument collections, libraries and other sources. Has an extensive bibliography.

551. Maddison, Francis. "Medieval Scientific Instruments and the Development of Navigational Instruments in the XVth and XVIth Centuries." Revista da Universidade de Coimbra, 24 (1971), 115-172.

Discusses medieval astronomical instruments known to the fifteenth century--the armillary sphere, the equatorium, the torquetum, astrolabe, quadrant, types of portable sundials, the mechanical clock and the compass. These instruments provided a tradition from which later navigational instruments developed.

552. Michel, Henri. "Un traité de l'astrolabe du XVe siècle." Homenaje à Millás Vallicrosa, II. Barcelona: Consejo superior de investigacions científicas, 1956, pp. 41-71.

Describes an astrolabe (Latin text included) from the early fifteenth century in a MS at the Bibliothèque Royale de Belgique, Brussels.

553. Millás Vallicrosa, J.M. "La introducción del cuadrante con cursor en Europa." Isis, 17 (1932), 218-258.

Traces the earliest appearance in the West of the quadrans vetus, the old quadrant with a cursor, a sliding plate, on the limb. The instrument was of Arabic origin. Discusses the relationship of the old quadrant with an earlier version (quadrans vetustissimus). Included in item 212.

554. Millás Vallicrosa, J.M. "Un nuevo tratado de astrolabio de R. Abraham ibn 'Ezra." al-Andalus, 5 (1940), 1-29.

Reports the discovery of a Latin work on the astrolabe by ibn 'Ezra in two MSS at the British Library. The treatise's emphasis on astrology, its treatment of the divergent opinions of different astronomers and schools and its similarity to ibn 'Ezra's Hebrew astrolabe work, attest to his authorship. Gives the Latin text.

555. Nielsen, Axel V. "A Sundial from about 1200 A.D." Centaurus, 12 (1968), 303-304.

Describes a wall dial on a church in Jutland. Argues that the displacement of the letters indicating the canonical hours can be attributed to an attempt to compensate for the deviation of the wall from true south.

556. North, John D. "Opus quorundam rotarum mirabilium." Physis, 8 (1966), 337-372.

Discusses a text from the first years of the fourteenth century or earlier which provides evidence of the design of astronomical clocks of a complex sort previously only known from general descriptions.

557. North, John D. "Monasticism and the First Mechanical Clocks." The Study of Time, II. Edited by J.T. Fraser and N. Lawrence. Berlin: Springer, 1975, pp. 381-398.

Stresses the role of the church in the development of clock mechanisms leading to the appearance in the thirteenth century of those with a mechanical escapement. Describes the astronomical clock designed for St. Albans by its abbot, Richard of Wallingford. See item 216.

558. Pedersen, Olaf. "The Life and Work of Peter Nightingale." Vistas in Astronomy, 9. Edited by Arthur Beer. Oxford: Pergamon Press, 1967, pp. 3-10.

Provides a reconstruction of the semissis of Petrus Philomena de Dacia who flourished 1291-1303. The instrument is a type of equatorium.

559. Peter of Maricourt. Il trattato dell'astrolabio di Pietro Peregrino di Maricourt. Edited by G. Boffito and C. Melzi d'Eril. Florence: Collegio alla Querce, 1927.

560. Poulle, Emmanuel. "L'astrolabe médiéval d'après les manuscrits de la Bibliothèque Nationale." Bibliothèque de l'ecole des chartes, 112 (1954), 81-103.

Describes astrolabe treatises and the work of prominent astrolabists based on MSS at Paris. Points out that medieval works on the astrolabe can be divided into those which describe its composition, those which list its uses and those on stereographic projection, although there are treatises which combine two or all of these features. After the thirteenth century, the astrolabe had assumed a standard form, perhaps due to the influence of Messahalla's popular text, although variations did appear. Lists the MSS in the Bibliothèque Nationale containing astrolabe treatises.

561. Poulle, Emmanuel. "La fabrication des astrolabes au moyen âge." Techniques et civilisations, 4 (1955), 117-128.

Draws on medieval texts to discuss the design and techniques of construction of the instrument.

562. Poulle, Emmanuel. "Peut-on dater les astrolabes médiévaux?" Revue d'histoire des sciences, 9 (1956), 301-322.

Argues that it is difficult to date the available medieval astrolabes from the instruments themselves. But astrolabe treatises contain precise astronomical information and can often be dated exactly. In examining several such treatises with reference to the longitude of the perihelion, the date of the vernal point and the coordinates of stars, concludes, however, that attempts to establish dates from this investigation are disappointing and would lead one to believe that the scientific value of the medieval astrolabe was debatable. However, the instrument, with its representation of the heavens, was an important teaching tool.

563. Poulle, Emmanuel. "L'équatoire de Guillaume Gilliszoon de Wissekerke," Physis, 3 (1961), 223-251.

Describes a cardboard equatorium in a MS in the Bibliothèque Nationale at Paris from the late fifteenth century by William of Carpentras who is identified as William of Wissekerke, originally from Zeeland. William's equatorium reduces the number of "instruments" for finding true longitudes to four, and does not require auxillary tables. The treatise is primarily devoted to the use of the instrument; it has a special "instrument" for precession, and also includes an astrolabium physicum, a modified astrolabe for medical-astrological purposes.

564. Poulle, Emmanuel. *Un constructeur d'instruments astronomiques au XVe siècle*: *Jean Fusoris*. Paris: Champion, 1963. Pp. 208.

Discusses the life, career and work of this fifteenth-century maker of astronomical instruments and composer of technical treatises in Latin and French. Treats Fusoris as an astrolabist, as a designer of an *equatorium* and as a maker of astronomical clocks, sundials and other instruments. Also discusses Fusoris' trigonometric tables.

565. Poulle, Emmanuel. "Bernard de Verdun et le *turquet*." *Isis*, 55 (1964), 200-208.

Points out that the *torquetum* described in the last tract of Bernard's *Tractatus super totam astrologiam* corresponds to that of Franco de Polonia, dated 1284. Argues that it is not possible to assign priority to either author, but that Bernard's is more likely. Gives the text of Franco's composition of the *torquetum*. See item 436.

566. Poulle, Emmanuel. "Le quadrant nouveau médiéval." *Journal des savants*, (1964), 148-167; 182-214.

Discusses treatises on and examples of the uses of the quadrant. These do not differ from that of the astrolabe, and like the latter, the new quadrant was not really employed for observational purposes. The instrument was invented by the Jewish astronomer and translator Profatius, who wished to improve the astrolabe by increasing its size which he did by reducing the face to one of its quadrants. The qualification "new" distinguishes the instrument from other quadrants, especially the old quadrant with cursor.

567. Poulle, Emmanuel. "Le traité d'astrolabe de Raymond de Marseille." *Studi medievali*, 5 (1964), 866-900.

Identifies Raymond's treatise in a MS at the Bibliothèque Nationale at Paris. Written in the first half of the twelfth century, but before 1141, it is perhaps the first independent work on the astrolabe in the Latin West. The originality of the treatise lies more in its literary style and in Raymond's expressed enthusiasm for astronomy than in its technical content.

568. Poulle, Emmanuel. "Théorie des planètes et trigonométrie au XVe siècle, d'après un _équatoire_ inedit, le _sexagenarium_." _Journal des savants_, (1966), 129-161.

Describes a type of little known _equatorium_--there is an excellent brass example at the Museum of the History of Science at Oxford. The reverse face of the Oxford _sexagenarium_ has a trigonometric network divided sexagesimally; hence the name. The purpose of the instrument is to reproduce by geometry the trigonometric formulas applicable to the astronomy of the prime mobile and to finding the true place of the planets.

569. Poulle, Emmanuel. "Un instrument astronomique dans l'occident latin, la _saphea_." _Studi medievali_, 10 (1969), 491-510.

Describes and compares several _saphea_ treatises. The instrument was a type of astrolabe which, however, employed a different principle of stereographic projection than the astrolabe. It was introduced to the West in 1231 and was the subject of numerous Latin treatises through the sixteenth century.

570. Poulle, Emmanuel. "Les instruments astronomiques de l'Occident latin aux Xe-XIIe siècles." _Cahiers de civilisation médiévale_, 15 (1972), 27-40.

Traces the history of astronomical instruments in the West, especially the astrolabe and the quadrant, beginning at the end of the tenth century with the first contacts with Arabic science.

571. Poulle, Emmanuel. "Les _équatoires_, instruments de la théorie des planètes au moyen âge." _Colloquia Copernicana_, 3. Warsaw: Polish Academy of Sciences, 1975, pp. 97-112.

Describes the _equatorium_, an instrument designed to find the true longitude of the planets without long calculation. _Equatoria_ are divided into three classes: geometrical (such as that of Campanus of Novara), arithmetical (for example, the _Albion_ of Richard of Wallingford) and trigonometrical, a class represented by one instrument, the _sexagenarium_. See item 572.

572. Poulle, Emmanuel. Les instruments de la théorie des planètes selon Ptolémée: Equatoires et horologerie planétaire du XIIIe au XVIe siècle. 2 vols. Paris: Champion, 1980. Pp. 1162.

The definitive work on these instruments representing two decades of investigation.

573. Price, Derek J. de Solla, ed. The Equatorie of the Planetis. Cambridge: Cambridge University Press, 1955. Pp. xvi + 214.

An edition with translation of a Middle English treatise composed around 1392 describing an equatorium which has features of both the instrument at Merton College and that described by John of Linières. The work was derived from a Latin version of an Arabic treatise; the accompanying tables are drawn from the Alfonsine Tables. The author is unknown, but there is evidence that the text may be by Chaucer and possibly a Chaucer holograph.

574. Price, Derek J. de Solla. "The Little Ship of Venice--A Middle English Instrument Tract." Journal of the History of Medicine and Allied Sciences, 15 (1960), 399-407.

Describes this portable sundial which tells time from the sun's altitude and is independent of latitude. The device is of Islamic origin and is described in many fourteenth-century MSS.

575. Tannery, Paul. "Le Traité du quadrant de Maître Robert Anglès, Montpellier XIIIe siècle." Mémoires scientifiques, 5. Issued by J.-L. Heiberg. Paris: Gauthier Villars, 1922, pp. 118-197.

Text of the treatise, here ascribed to Robert Anglicus.

576. Taylor, F. Sherwood. "Mediaeval Scientific Instruments." Discovery, 11 (1950), 282-287.

Surveys instruments used in time-telling such as the astrolabe, sundials and mechanical clocks. Also discusses the compass, instruments for optical investigations, alchemical apparatus and medical instruments.

577. Thorndike, Lynn. "Of the Cylinder Called the Horologe of Travelers." Isis, 13 (1929-30), 51-52.

Reports a Vatican MS version of this cylindrical portable sundial known as the traveler's dial which was introduced to the Latin West by Hermannus Contractus in the eleventh century.

578. Thorndike, Lynn. "Who Wrote Quadrans vetus?" Isis, 37 (1947), 150-153.

The work on the quadrans vetus (old quadrant) has been attributed to both Robertus Anglicus and John of Montpellier; most of the MSS materials, as well as learned tradition, support the claim of John.

579. Thorndike, Lynn. "Date of Peter of St. Omer's Revision of the New Quadrant of Profatius Judaeus." Isis, 51 (1960), 204-206.

Contends that the revision of Profatius' quadrant treatise by Peter was made earlier than the previously assigned date of 1309.

580. Tomba, Tullio. "Un astrolabio de XIV secolo di probabile origine Italiana." Physis, 12 (1970), 82-87.

Describes an astrolabe of Italian origin dating from the end of the fourteenth century which is in an excellent state of conservation.

581. Zinner, Ernst. Deutsche und niederländische astronomische Instrumente des 11.-18. Jahrhunderts. Munich: C.H. Beck, 1956. Pp. 679.

Surveys a wide range of instruments--sundials, clocks, planetaria, instruments for astrological purposes, for observation and field measurement. The larger portion of the book contains a list of instrument makers and authors of works on instruments.

# PHILOSOPHY, METAPHYSICS, METHODOLOGICAL CONSIDERATIONS, LOGIC

## PHILOSOPHY: GENERAL

582. Gilson, Étienne. History of Christian Philosophy in the Middle Ages. New York: Random House, Pp. xvii + 829.

A detailed study of the history of philosophical ideas from Justin Martyr to Nicholas of Cusa with extensive notes.

583. McKeon, Richard P., ed. Selections from Medieval Philosophers. 2 vols. New York: Charles Scribner's Sons, 1929, 1930. Pp. xx + 375, xviii + 515.

Vol. I: From Augustine to Albert the Great; Vol. II; From Roger Bacon to William of Ockham. Contains selections from the work of 15 philosophers centered around the problem of knowledge.

584. Maurer, Armand A. Medieval Philosophy. New York: Random House, 1962. Pp. xix + 435.

A survey from the age of the fathers through the Renaissance designed for the general reader and for undergraduates.

585. Vignaux, Paul. Philosophy in the Middle Ages: An Introduction. Translated by E.C. Hall. Westport, Conn.: Greenwood Press, 1973. Pp. 223.

A translation of Philosophie au moyen âge, 1959. Surveys philosophical and theological thought from the Carolingian Renaissance through the contributions of the chief scholastics of the thirteenth and fourteenth centuries.

586. Weinberg, Julius R. A Short History of Medieval Philosophy. Princeton: Princeton University Press, 1964. Pp. x + 340,

Sketches philosophical thought from St. Augustine through the fourteenth century, including Islamic and Jewish philosophy; designed for the general reader.

## PHILOSOPHY AND METAPHYSICS: SPECIFIC TOPICS

587. Aegidius Romanus. Giles of Rome: Errores philosophorum. Edited with critical text, notes and introduction by Joseph Koch. Translated by John O. Riedl. Milwaukee: Marquette University Press, 1944. Pp. lix + 67.

An edition with English translation and an analysis of the sources of Aegidius Romanus' Errores philosophorum Aristotelis, Averrois, Avicennae, Algazalis, Alkindi, Rabbi Moysis. To Aegidius, the philosophical doctrines of these writers presented a danger to Christian orthodoxy.

588. Aegidius Romanus. Questiones methaphisicales. Venice, 1501. Reprint. Frankfort: Minerva, 1966.

589. Aegidius Romanus. Super librum de causis. Venice, 1550. Reprint. Frankfort: Minerva, 1968.

590. Algazel (al-Ghazzālī). Algazel's Metaphysics: A Medieval Translation. Edited by J.T. Muckle. Toronto: St. Michael's College, 1933. Pp. xix + 247.

591. Alkindi (al-Kindī). Die philosophischen Abhandlungen des Ja'qūb ben Isḥāq. Edited by A. Nagy. Beiträge zur Geschichte der Philosophie des Mittelalters, 2. Münster: Aschendorff, 1897. Pp. xxxiv + 84.

Contains the De intellectu, De somno et visione, De quinque essentiis, and the Introductorius in artem logicae demonstrationis.

592. Aquinas, Thomas. In duodecim libros metaphysicorum Aristotelis expositio. Edited by M.R. Cathala and R. M. Spiazzi. Turin: Marietti, 1950. Pp. xxiii + 647.

593. [pseud.] Aristotle. De mundo. Translationes Bartholomaei et Nicholai. Edited by William L. Lorimer, revised by Lorenzo Minio-Paluello. Aristoteles latinus, XI, 1-2. Bruges: Desclée de Brouwer, 1965. Pp. iv + 194.

594. Aristotle. Metaphysica, I-IV. 4. Translatio Iacobi sive "vetustissima" cum scholiis et translatio composita sive "vetus." Edited by Gudrun Vuillemin-Diem. Aristoteles latinus, XXV, 1-1a. Brussels: Desclée de Brouwer, 1970. Pp. lix + 244.

595. Aristotle. Metaphysica, I-X, XII-XIV. Translatio anonyma sive "media." Edited by Gudrun Vuillemin-Diem. Aristoteles latinus XXV, 2. Leiden: Brill, 1976. Pp. lxix + 385.

596. Averroes (Ibn Rushd). Die Epitome der Metaphysik des Averroes. Translated with an introduction and notes by Simon van den Bergh. Leiden: Brill, 1970. Pp. xxv + 329.

597. Avicebron (Ibn Gabirol). Fons vitae. Edited by Clemens Baeumker. Beiträge zur Geschichte der Philosophie des Mittelalters, 1, 2-4. Münster: Aschendorff, 1895. Pp. xxii + 558.

598. Avicenna (Ibn Sina). Metaphysica sive prima philosophia. Venice, 1495. Reprint. Frankfort: Minerva, 1966.

599. Avicenna (Ibn Sina). Liber de philosophia prima sive Scientia divina. Edition of the medieval Latin translation by Simone van Riet with introduction by Gerard Verbeke. Louvain: E. Peeters, 1977--.

Latin text and French translation of the fourth part of the al-Shifā.

600. Baeumker, Clemens. Witelo, ein Philosoph und Naturforscher des XIII. Jahrhunderts. Beiträge zur Geschichte der Philosophie des Mittelalters, 3. Münster: Aschendorff, 1908. Pp. xxii + 686.

An edition of the De intelligentiis with a discussion of sources and analysis.

601. Baeumker, Clemens, ed. "Das pseudo-hermetische Buch der vierundzwanzig Meister (Liber XXIV philosophorum): Ein Beitrag zur Geschichte des Neupythagoreismus und Neuplatonismus im Mittelalter." Gesammelte Vorträge und Aufsätze, 6. Beiträge zur Geschichte der Philosophie des Mittelalters, 25, 1. Münster: Aschendorff, 1928, pp. 194-214.

602. Bardenhewer, Otto, ed. Die pseudo-aristotelische Schrift über das reine Gute bekannt unter dem Namen Liber de causis. Freiburg im Breisgau, 1882. Reprint. Frankfort: Minerva, 1957? Pp. xviii + 330.

A reworking of the Elementatio theologica of Proclus translated from Arabic by Gerard of Cremona.

603. Bettoni, Efrem. Duns Scotus: The Basic Principles of his Philosophy. Translated by Bernardine Bonansea. Washington, D.C.: Catholic University of America Press, 1961. Pp. 220.

A survey of Scotus' philosophical positions.

604. Birkenmajer, Aleksander."Études sur Witelo, I-IV." Bulletin international de la Academie Polonaise des Sciences. Classe de philologie. Classe d'histoire et de philosophie, 1918 (pub. 1920), 4-6; 1919-20 (pub. 1925), 354-360; 1922 (pub. 1925), 6-9.

Summaries of papers given at the Academy on various aspects of Witelo's work, thought and career. Discusses a MS containing extracts from his De natura daemonum, rejects his authorship of the De intelligentiis, investigates the Arabic-neoplatonic sources of Witelo's metaphysics and treats his stay at Padua.

605. Boehner, Philotheus. "The notitia intuitiva of Non-Existents according to William of Ockham." Traditio, 1 (1943), 223-275.

Argues from an analysis of Ockham's Sentences commentary that his interpretation of the intuitive perception of non-existent objects was not an avenue to scepticism. While the deity could produce a supernatural knowledge of an object which did not actually exist, according to Ockham, a perfect intuitive knowledge whether of existent or non-existent things is infallible knowledge.

606. Boehner, Philotheus. Collected Articles on Ockham. Edited by E.M. Buytaert. St. Bonaventure, N.Y.: Franciscan Institute, 1958.

Twenty-four articles by Boehner, ordered according to topic, with notes and cross-references on Ockham research in general, and on themes in philosophy, logic, epistemology, metaphysics, theology and politics. Includes item 605.

607. Boetius of Dacia. Boetii de Dacia Tractatus de aeternitate mundi. Edited by Géza Sajó. Berlin: de Gruyter, 1964. Pp. 70.

608. Buridan, John. In metaphysicen Aristotelis. Paris, 1588. Reprint. Frankfort: Minerva, 1964.

609. Chenu, M.-D. Nature, Man, and Society in the Twelfth Century. Selected, edited and translated by Jerome Taylor and Lester K. Little with a preface by Étienne Gilson. Chicago: University of Chicago Press, 1968.

Nine essays plus the introduction and the preface by Gilson taken from La théologie au douzième siècle, 1957 which, using theology as a point of departure, discuss the cultural and intellectual movements of a pivotal century.

610. Cusa, Nicholas of. Of Learned Ignorance. Translated by Germain Heron with introduction by D.J.B. Hawkins. London: Routledge and Kegan Paul, 1954. Pp. xxviii + 174.

611. Cusa, Nicholas of. Opere filosofiche. Edited by Graziella Federici Vescovini. Turin: Unione tipografico-editrice torinese, 1972. Pp. 1069.

611a. Dales, Richard C. "Discussions of the Eternity of the World during the First Half of the Twelfth Century." Speculum, 57 (1982), 495-508.

Points out that the topic of the eternity of the world, a source of marked conflict between Greek philosophy and Christian dogma, was much discussed in the first half of the twelfth century, inspired by the Timaeus rather, as in the subsequent century, the ideas of Aristotle. While there was no universal agreement, these twelfth-century arguments did not occasion opposition.

612. Gilson, Étienne. "Avicenne en occident au moyen âge." Archives d'histoire doctrinale et littéraire du moyen âge, 36 (1969), 89-121.

Maintains that Avicenna's impact on the West can be divided into three phases--an initial period of influence, a decline after the arrival of Aristotle followed by a revival. Emphasizes that one cannot speak of a Latin "Avicennism" parallel to Latin Averroism. Averroes separated the domains of theology and philosophy, but Avicenna never disengaged his philosophical speculations from his Moslem faith.

613. Henry of Ghent. Les Quaestiones in librum de causis attribuées à Henri de Gand. Edited by John P. Zwaenepoel. Louvain: Publications universitaires, 1974. Pp. 160.

614. Hermann of Carinthia. De essentiis. Edited by Manuel Alonso Alonzo. Comillas (Santander): Universidad Pontifica, 1946. Pp. 104.

The text of Hermann's philosophical treatise can also be found in item 122, pp. 48-49.

614a. John of Jandun. Quaestiones in duodecim libros metaphysicae. Venice, 1553. Reprint. Frankfort: Minerva, 1966.

615. Klibansky, Raymond. The Continuity of the Platonic Tradition During the Middle Ages. London: Warburg Institute, 1939. Pp. 58.

Traces the main currents of medieval Platonism stemming from Byzantine, Arabic and Latin traditions.

616. Krebs, Engelbert. Meister Dietrich, Theodoricus Teutonicus de Vriberg: Sein Leben, seine Werke, seine Wissenschaft. Beiträge zur Geschichte der Philosophie des Mittelalters, 5. Münster: Aschendorff, 1906. Pp. xii + 155 + 230.

Provides excerpts of and commentary on Theodoric's works on logic, optics, natural philosophy and theology, plus texts of the De intellectu et intelligibili and De habitibus.

617. Kuksewicz, Zdzislaw. "The Problem of Walter Burley's Averroism." Studi sul XIV secolo in memoria Anneliese Maier. Edited by Alfonso Maierù and Agostino Paravicini Bagliani. Rome: Edizioni di storia e letteratura, 1981, pp. 341-377.

Argues that Burley's position vis-à-vis Averroism went through an evolution from rejection to acceptance. By the writing of Book II of his commentary on De anima, he reveals a rather accepting attitude which, in Book III, had firmed sufficiently to mark him as an Averroist.

618. McClintock, Stuart. Perversity and Error: Studies on the "Averroist" John of Jandun. Bloomington, Ind.: Indiana University Press, 1956. Pp. viii + 204.

Argues that John's reputation as "perverse and erroneous" is too harsh a judgment. Maintains that when John, following Averroes, arrived at conclusions where reason countered Christian dogma, he carefully stated that truth lay with faith. Nor was he insincere or hypocritical; his approach was not untypical of his contemporaries.

619. Maier, Anneliese. "Wilhelm von Alnwick's Bologneser Quaestionen gegen den Averroismus (1323)." Ausgehendes Mittelalter: Gesammelte Aufsätze zur Geistesgeschichte des 14. Jahrhunderts, I. Rome: Edizioni di storia e letteratura, 1964, pp. 1-40.

Discusses those questions disputed by Alnwick dealing with Averroist teaching on the unity of the intellect, the eternity of the world and the question whether God knows things other than himself, and directed toward Bolognese Averroists and the English Averroist Thomas of Wylton. Argues that the interpretation of the doctrine of the "double truth" altered in the course of the fourteenth century as did the notion of what constituted an "Averroist." Reprinted from Gregorianum, 30 (1949), 265-308.

620. Maier, Anneliese. "Die Bologneser Philosophen des 14. Jahrhunderts." Ausgehendes Mittelalter: Gesammelte Aufsätze zur Geistesgeschichte des 14. Jahrhunderts,

II. Rome: Edizioni di storia e letteratura, 1967, pp. 335-349.

Confirms by a study of new material and of texts previously known that the Bologna arts faculty was an important Averroist center in the early fourteenth century. Reprinted from Studi e memorie per la storia dell' universitá di Bologna, nuova serie, 1 (1955), 297-310.

621. Maier, Anneliese. "Ein unbeachteter 'Averroist' des 14. Jahrhunderts: Walter Burley." Ausgehendes Mittelalter: Gesammelte Aufsätze zur Geistesgeschichte des 14. Jahrhunderts, I. Rome: Edizioni di storia e letteratura, 1964, pp. 101-121.

Points out that scholastics could interpret Averroes' defense of the unity of the intellect as representing Aristotle's own opinion. In this sense, Burley was an Averroist. Also, in his exposition on the De anima, Burley appears to have had some sympathy for the Averroist position. If so, this may cast Burley's known connection with Bolognese Averroist arts mästers in a new light. Reprinted from Medioevo et rinascimento: Studi in onore di Bruno Nardi, Florence, 1955, 476-499.

622. Maier, Anneliese. "Das Problem der Evidenz in der Philosophie des 14. Jahrhunderts." Ausgehendes Mittelalter: Gesammelte Aufsätze zur Geistesgeschichte des 14. Jahrhunderts, II. Rome: Edizioni di storia e letteratura, 1967, pp. 367-418.

Argues that the basis of medieval scepticism or uncertainty as to knowledge lies in the position taken by Ockham in his Quodlibeta that God could cause an apparent evident assent to a non-existent which could not be distinguished from the genuine. However, while the supernatural possibility of God's interference in the natural order was universally accepted, there was an optimistic conviction that reality could be grasped through the senses which was more characteristic of the fourteenth century than scepticism. See item 605.

623. Maimonides, Moses. The Guide of the Perplexed. Translated with an introduction and notes by Shlomo Pines. Chicago: University of Chicago Press, 1963. Pp. cxxxiv + 658.

624. Mandonnet, Pierre. Siger de Brabant et l'averroism latin au XIIIe siècle. 2d ed. Louvain: Institute supérieur de philosophie de l'université, 1908, 1911. Pp. xvi + 328, xxxi + 189.

Discusses in Part I the effect of the reception of Aristotle-Averroes at Paris and the doctrinal agitation leading to the first condemnations of 1270, with emphasis on Siger's identification as a Latin Averroist, his literary activity and an evaluation of his role. Outlines the continuation of the troubles at Paris resulting in the condemnations of 1277 and Siger's final years after the crisis. Part II contains texts, mostly treatises by Siger, as well as a list of the 1277 condemnations arranged by theme.

625. Moody, Ernest A. "Empiricism and Metaphysics in Medieval Philosophy." Studies in Medieval Philosophy, Science, and Logic: Collected Papers, 1933-1969. Berkeley: University of California Press, 1975, pp. 287-304.

Argues that the rise of empiricism and decline of metaphysics in late medieval thought is a decline in philosophy only if the goal of philosophy is considered as the establishment of a metaphysics designed to support theology. A philosophical empiricism arose from the long-standing anti-metaphysical attitude of the church which, fostered by Ockham and the aftermath of 1277, directed theologians toward a critique of metaphysics and toward positive science.

626. Moody, Ernest A. "Ockham, Buridan, and Nicholas of Autrecourt." Studies in Medieval Philosophy, Science, and Logic: Collected Papers, 1933-1969. Berkeley: University of California Press, 1975, pp. 127-160.

Argues that the 1340 statute of condemnation at Paris was not addressed to Ockham and Ockhamists, but to Nicholas of Autrecourt, who was unsympathetic to Ockham's views and who did hold a position specifically attacked in 1340. Buridan, a moderate Ockhamist and a critic of Nicholas, signed the statute as rector at Paris. The statement in the 1340 statute drawing attention to the one of 1339 (which was anti-Ockham) can be interpreted as a warning that the earlier prohibitions were still in effect.

627. Moody, Ernest A. "A Quodlibetal Question of Robert Holkot, O.P., on the Problem of the Objects of Knowledge and Belief." Studies in Medieval Philosophy, Science, and Logic: Collected Papers, 1933-1969. Berkeley: University of California Press, 1975, pp. 321-352.

Points out that Holkot's quodlibetal question of 1332 presents two major positions on this problem: that of Ockham, who held that the proposition was the object of knowledge, and that of his opponent, Walter Chatton, who maintained it to be the thing signified by the proposition.

628. Moody, Ernest A. Studies in Medieval Philosophy, Science, and Logic: Collected Papers, 1933-1969. Berkeley: University of California Press, 1975. Pp. xix + 453.

A collection of previously published articles. Includes items 254, 328, 329, 330, 625, 626, 627, 679 and 1124.

629. Paqué, Ruprecht. Das Pariser Nominalistenstatut zur Entstehung des Realitätsbergriffs der neuzeitlichen Naturwissenschaft (Occam, Buridan und Petrus Hispanus, Nikolaus von Autrecourt und Gregor von Rimini). Berlin: de Gruyter, 1970. Pp. viii + 337.

Argues from an analysis of its sections and of the issues which provoked it that the 1340 "nominalists statute" at Paris was directed at Ockham and "Ockhamists" rather than at Nicolas of Autrecourt, as Moody has suggested. Discusses the role of Buridan, rector at Paris when the statute was promulgated, and his stance vis-à-vis Ockhamist thought. See item 626.

630. Scott, Jr., T.K. "John Buridan on the Objects of Demonstrative Science." Speculum, 40 (1965), 654-673.

Discusses the position of William of Ockham, Gregory of Rimini and Jean Buridan on whether the objects of knowledge are eternal if the propositions of demonstrable science are true.

630a. Scott, T.K. "Ockham on Evidence, Necessity, and Intuition." Journal of the History of Philosophy, 7 (1969), 27-49.

Examines these three notions, establishing the interrelation among them and showing both how they operate in Ockham's theory of science and how his interpretation of their relationship might have led to the critical and sceptical orientation of the fourteenth century, tendencies attributed to Ockham.

631. Scott, Jr., T.K. "Nicholas of Autrecourt, Buridan and Ockhamism." Journal of the History of Philosophy, 9 (1971), 15-41.

Refutes the notion that Buridan's theory of knowledge is dependent on Ockham, an assumption which is based on Buridan's defense of Ockham's views against the scepticism of Nicholas. Concludes that neither Autrecourt nor Buridan were Ockhamists, although Nicholas' position is closer to Ockham than is that of Buridan.

632. Shapiro, Herman. "A Note on Burley's Exaggerated Realism." Franciscan Studies, 20 (1960), 205-214.

Argues that while Burley's reputation as an independent thinker has been vindicated, he is still classified as an exaggerated realist. From an analysis of Burley's attitude toward universals in his De materia et forma, concludes that Burley should be seen as a conceptualist, a compromise position between extreme realism (universals have an extra-mental existence, which Burley denies), and nominalism (no universality in the mind or in things). To Burley, universals had a psychological being in the mind.

633. Shapiro, Herman. "More on the 'Exaggeration' of Burley's Realism." Manuscripta, 6 (1962), 94-98.

Argues that Burley's reputation as an extreme realist is not justified, and that at some period, Burley changed his viewpoint to one of moderate realism. The small treatise edited here, beginning Notandum quod tantum sunt, clarifies this change. See item 632.

634. Sharp, D.E. Franciscan Philosophy at Oxford in the Thirteenth Century. Oxford: Oxford University Press, 1930. Pp. viii + 419.

Discusses the philosophical speculations of prominent Oxford Franciscans (including Grosseteste): Thomas of York, Bacon, Pecham, Richard of Middleton and Duns Scotus.

635. Siger of Brabant. Die Impossibilia des Siger von Brabant, eine philosophische Streitschrift aus dem XIII. Jahrhundert. Edited by Clemens Baeumker. Beiträge zur Geschichte der Philosophie des Mittelalters, 2, 6. Münster: Aschendorff, 1898. Pp. viii + 200.

636. Siger of Brabant. Questions sur la Metaphysique. Edited by C.A. Graiff. Louvain: Institut supérieur de philosophie, 1948. Pp. xxxii + 399.

637. Siger of Brabant. Les Questiones super librum de causis. Edited by Antonio Marlasca. Louvain: Publications universitaires, 1972. Pp. 211.

638. van Steenberghen, Fernand. "La philosophie de la nature au XIIIe siècle." La filosofia della natura nel medioevo. Atti del terzo congresso internazionale di filosofia medioevale, 1964. Milan: Vita e pensiero, 1966, pp. 114-132.

Argues that the philosophy of nature of this century was dominated by Aristotle although the Greek master's thought was inserted in a metaphysical system based on Arabic and Christian Neoplatonism.

639. van Steenberghen, Fernand. Aristotle in the West: The Origins of Latin Aristotelianism. 2d ed. Translated by Leonard Johnston. Louvain: Nauwelaerts, 1970. Pp. 244.

An amplified version of the French edition of 1946. Traces the impact of Greek and Arabic thought on the Christian West, including the introduction and triumph of Aristotle at Paris, the Neoplatonizing Aristotelianism of Oxford, the rise of a heteorodoxy derived from Aristotle and the crisis which led to 1277.

640. van Steenberghen, Fernand. "La controverse sur l'éternité du monde au XIIIe siècle." Bulletin de l'Académie Royale de Belgique, classe des lettres et des sciences morales et politiques, sér. 5, 58 (1972), 267-287.

Delineates three attitudes vis-à-vis the eternity of the world arising from the confrontation of Aristotelian and Arabic thought with Christian doctrine. Bonaventure defended the traditional theological position, while Siger of Brabant advocated the world's eternity at one stage of his career. Aquinas represented a compromise position stating that the origin of the world in time could not be demonstrated, but must be assumed true based on faith and revelation.

641. van Steenberghen, Fernand. Maître Siger de Brabant. Louvain: Publications universitaires, 1977. Pp. 442.

Argues that before 1270, Siger was a radical Aristotelian with strong Averroist leanings. After 1270, his thought, although still heterodox, reflects an inner crisis, a weakened adherence to heretical doctrine and the growing influence of Aquinas.

642. Vanni Rovighi, Sofia. "Gli Averroisti bolognesi." Oriente et occidente nel medioevo: Filosofia e scienze. Rome: Accademia nazionale dei Lincei, 1971, pp. 161-179.

Argues that there was a flourishing group of Averroists at Bologna in the second decade of the fourteenth century where Averroist doctrines condemned at Paris in 1270 and 1277 were disputed and discussed.

643. Vollert, Cyril, Lottie H. Kendzierski and Paul M. Byrne, trans. St. Thomas Aquinas, Siger of Brabant, St. Bonaventure: On the Eternity of the World (De aeternitate mundi). Milwaukee: Marquette University Press, 1964. Pp. xii + 117.

English translations of works of Aquinas and Siger, as well as selections from Bonaventure, on this topic.

644. Weisheipl, James A. "Albertus Magnus and the Oxford Platonists." Proceedings of the American Catholic Philosophical Association, 32 (1958), 124-139.

Argues that Albert's attack on Platonic notions in his paraphrase of the Metaphysics was directed against contemporary amici Platonis at Oxford: Grosseteste, Bacon, and particularly Robert Kilwardby, whose opinion of the status of mathematics in a hierarchy of learning and as a fundamental principle of reality Albert opposed. To Albert, far from being the basis of nature, mathematics was merely a "remote aid" to its study.

645. Wippel, John F. "The Condemnations of 1270 and 1277 at Paris." Journal of Medieval and Renaissance Studies, 7 (1977), 169-201.

Analyses the background which led up to the late thirteenth-century crisis engendered by the reception of Greek philosophical thought in a Christian framework. Discusses the nature of several of the condemned theses, toward whom they were addressed and the subsequent effect of the condemnations. Argues that they had a deleterious effect on Thomism, encouraged the growth of neo-Augustinian thought and created an antagonism between theology and philosophy.

646. Wippel, John F. "Did Thomas Aquinas Defend the Possibility of an Eternally Created World?" (The De aeternitate mundi Revisited)." Journal of the History of Philosophy, 19 (1981), 21-37.

Argues that in his De aeternitate mundi, Aquinas took the position that the creation of the world in time could not be demonstrated and that an eternally created world was possible.

647. Wolter, Allan, ed. and trans. Duns Scotus, Philosophical Writings: A Selection. London: Thomas Nelson, 1962. Pp. xxiii + 198.

Selections from Scotus on metaphysics, the nature of God and human knowledge.

648. Würsdörfer, Joseph. "Erkennen und Wissen nach Gregor von Rimini: Ein Beitrag zur Geschichte der Erkenntnistheorie des Nominalismus." Beiträge zur Geschichte der Philosophie des Mittelalters, 20 (1917), 1-139.

Treats Gregory's theory of knowledge in the light of the author's conviction that Gregory was a disciple of William of Ockham.

## METHODOLOGICAL CONSIDERATIONS

649. Ashley, Benedict M. "St. Albert and the Nature of Natural Science." Albertus Magnus and the Sciences: Commemorative Essays. Edited by James A. Weisheipl. Toronto: Pontifical Institute of Mediaeval Studies, 1980, pp. 73-102.

Argues that Albert had a well-developed scientific methodology, essentially Aristotelian, with remanents of Neoplatonism, logically organized to investigate physical bodies and the causes of the changes they undergo in natural processes.

650. Bolzan, J.E. "¿Navaja de Ockham o navaja de Santo Tomas?" Sapientia, 29 (1974), 207-216.

Argues that the principle of economy of the Posterior Analytics was used in some form by numerous authors from late antiquity on. Particularly, the principle, although identified with Ockham (Ockham's razor), was employed, directly or indirectly, by Aquinas.

651. Crombie, A.C. Robert Grosseteste and the Origins of Experimental Science, 1100-1700. Oxford: Clarendon Press, 1953. Pp. ix + 369.

Argues that the modern notion of the experimental method was, at least qualitatively, created by scholastics in thirteenth-century Europe. The first indications of the characteristics of this methodology can be seen in Grosseteste and his followers at Oxford.

652. Crombie, A.C. "Grosseteste's Position in the History of Science." Robert Grosseteste, Scholar and Bishop. Edited by D.A. Callus. Oxford: Clarendon Press, 1955, pp. 98-120.

Outlines Grosseteste's innovations in scientific method based on the newly-introduced Posterior Analytics, his pioneering stress on the value of experiment and the high value he gave to mathematics. His influence at Oxford, which was to develop a tradition of the application of mathematics to natural philosophy, was significant.

653. Crombie, A.C. "The Significance of Medieval Discussions of Scientific Method for the Scientific Revolution." Critical Problems in the History of Science. Edited by Marshall Clagett. Madison: University of Wisconsin Press, 1959, pp. 79-101.

Finds evidence in the thirteenth and fourteenth centuries of a conception of the purpose of science and its methods which links the late Middle Ages and the seventeenth century in a continuity of a scientific tradition which began with the Greeks.

654. Crombie, A.C. "Quantification in Medieval Physics." Isis, 52 (1961), 143-160.

Argues that while scholastics used quantified concepts to explain theoretical ideas, they did not quantify the results of experiments or observations, an attitude toward method furthered by university learning. On occasion, practical considerations did lead to the application of quantification to theoretical concepts, but primarily among craftsmen who had little connection with the scholastic milieu.

655. Easton, Stewart C. Roger Bacon and his Search for a Universal Science. Oxford: Blackwell, 1952. Pp. vii + 255.

An intellectual biography of Bacon which focuses on his vision of a unified philosophy which would encompass all of the sciences.

656. Eastwood, Bruce S. "Medieval Empiricism: The Case of Grosseteste's Optics." Speculum, 43 (1968), 306-321.

Analyzes Grosseteste's optical treatises for evidence of his scientific methodology and the position he ascribed to experiment. Concludes that Grosseteste's estimation of the value of empirical or experimental methods was qualified.

657. Evans, Gillian R. "More geometrico: The Place of the Axiomatic Method in the Twelfth-Century Commentaries on Boethius' Opuscula sacra." Archives internationales d'histoire des sciences, 27 (1977), 207-221.

Argues that twelfth-century students engaged with Boethius' theological works reveal a tendency toward a "Euclidean" or axiomatic approach. However, their use of this methodology was limited as the problems they addressed did not lend themselves to numerical treatment.

658. Fisher, N.W. and Sabetai Unguru. "Experimental Science and Mathematics in Roger Bacon's Thought." Traditio, 27 (1971), 353-378.

Argues that Bacon's achievements lay not in specific applications and original discoveries in mathematics and natural philosophy, but in his high estimation of the uses of mathematics and experiment in the teaching and pursuit of science.

659. Goldstein, Bernard R. "The Status of Models in Ancient and Medieval Astronomy." Centaurus, 24 (1980), 132-147.

Traces the use of geometrical models employed to account for celestial motions. Beginning in antiquity, these models introduced a long-standing confrontation between physical and mathematical astronomy as to whether a model should correspond to physical reality or simply account for the phenomena. In the medieval period, various responses arose, particularly among Arabic and Jewish astronomers who devised mathematical models that harmonized with accepted physical principles.

660. Grabmann, Martin. Die Geschichte der scholastischen Methode. 2 vols. Freiburg in Breisgau, 1909, 1911. Reprint. Graz: Akademische Druck- u. Verlagsanstalt, 1957. Pp. xiii + 354, vii + 586.

Defines scholastic method as the application of reason, i.e., philosophy to the truths of revelation. The pursuit of this goal did eventually produce a technique used in university instruction and as a literary form (the questio-format). Traces the development of this method from the Church Fathers and Boethius through its fulfillment in the twelfth and early thirteenth centuries as revealed in the Sentences-literature and the thought of several scholastics (Abelard, Hugh of St. Victor, Robert of Melun and the Chartrians).

661. Grant, Edward. "Hypotheses in Late Medieval and Early Modern Science." Daedalus, 91 (1962), 599-612.

Stresses the difference in the medieval conception of a scientific hypothesis and that of the early modern period. To scholastics, an hypothesis did not need to represent physical reality so long as it accounted for the phenomena. This notion, stemming from ancient astronomy, re-enforced by the condemnations of 1277 and the ideas of Ockham, served to produce a limited confidence in human reason's ability to arrive at physical certainty. Copernicus marked a break with this attitude in his insistence that a hypothesis reflect physical reality.

662. Grant, Edward. "Late Medieval Thought, Copernicus and the Scientific Revolution." Journal of the History of Ideas, 23 (1962), 197-220.

Argues that there was a marked break between the late medieval conception of a hypothesis and that held by Copernicus. The notion that a hypothesis need only save the phenomena, familiar from Greek astronomy, re-enforced by the 1277 condemnations and Ockham's radical empiricism, created an atmosphere in which previous metaphysical suppositions were destroyed and hypotheses about nature matters of choice between alternatives. But, to Copernicus, a scientific hypothesis represented a truth which corresponded to physical reality.

663. Grant, Edward. "Aristotelianism and the Longevity of the Medieval World View." History of Science, 16 (1978), 93-106.

Argues that the Aristotelian cosmology and physical theory which made up the medieval world view was not susceptible to internal reform or comprehensive attack because the many scholastic disagreements and reinterpretations of details of its doctrines were not formulated in a general and synthetic fashion. Scholastic disputes on various details of the Aristotelian cosmos were treated in isolated questiones. The eventual collapse of Aristotelianism was due to the onslaught of new ideas that destroyed both the general cosmological framework and attacked the validity of the details of its operation.

664. Grant, Edward. "Scientific Thought in Fourteenth-Century Paris: Jean Buridan and Nicole Oresme." Machaut's World: Science and Art in the Fourteenth Century. Edited by Madeleine Pelner Cosman and Bruce Chandler. New York: New York Academy of Sciences, 1978, pp. 105-124.

Stresses the differing conceptions of the possibility and validity of scientific knowledge in the thought of these two scholastics. Both agreed that absolute truth was found only in faith, but the secular schoolman, Buridan assigned great importance to the effectiveness of human reason and experience in determining truths of nature. Oresme, however, denigrated the ability and power of human reason which could be confronted with alternative explanations where definitive truth could not be demonstrated.

665. Grant, Edward. "The Condemnation of 1277, God's Absolute Power, and Physical Thought in the Late Middle Ages." Viator, 10 (1979), 211-244.

Discusses the consequences of the theological reaction to the threat of Greek determinism, particularly as embodied in the new Aristotle, which culminated in the condemnations of 1277. The chief feature of this reaction was a strong emphasis on God's absolute power which provided fourteenth-century natural philosophers with a license to multiply imaginative hypothetical speculations completely foreign to Aristotelian doctrines, especially his strictures on an extracosmic void space. While these speculations did not mean the demise of Aristotle, those which introduced void space were to have far-reaching results.

666. Hackett, Jeremiah M.G. "The Attitude of Roger Bacon to the Scientia of Albertus Magnus." Albertus Magnus and the Sciences: Commemorative Essays. Edited by James A. Weisheipl. Toronto: Pontifical Institute of Mediaeval STudies, 1980, pp. 53-72.

Argues from a survey of Bacon's works that Albert may be identified with the "unknown master" whose conception of the nature of science Bacon deplored. Albert is known as having rejected the view that mathematics was basic to an explanation of nature; Bacon, with his emphasis on the mathematical foundation of reality, was opposed to Albert's position.

667. Molland, A.G. "Nicole Oresme and Scientific Progress." Antiqui und Moderni: Traditionsbewusstsein und Fortschrittsbewusstsein im späten Mittelalter. Edited by Albert Zimmermann. Berlin: de Gruyter, 1974, pp. 206-220.

Argues based on his De configurationibus and the De commensurabilitate that Oresme displayed considerable scepticism concerning the possibility of a certain knowledge of nature. For example, neither mathematical demonstration nor sense impression could determine whether the celestial motions were commensurable or not, although their incommensurability is more probable.

668. Molland, A.G. "Mathematics in the Thought of Albertus Magnus." Albertus Magnus and the Sciences: Commemorative Essays. Edited by James A. Weisheipl. Toronto: Pontifical institute of Mediaeval Studies, 1980, pp. 463-478.

Characterizes Albert as a conceptualist to whom mathematics was a mental activity which could not contribute much that was fundamental to an understanding of the external world. His position is revealed in his attacks on Pythagoreans and Platonists to whom mathematics provided the basis of nature.

669. Murdoch, John E. "Philosophy and the Enterprise of Science in the Later Middle Ages." The Interaction between Science and Philosophy. Edited by Yehuda Elkana. Atlantic Highlands, N.J.: Humanities Press, 1974, pp. 51-74.

Maintains that medieval science in the fourteenth century was frequently identical with medieval philosophy. The new conceptual languages which distinguished the period were applied to problems within a philosophical context by scholastics who were philosophers even though natural philosophers.

669a. Murdoch, John E. "From Social into Intellectual Factors: An Aspect of the Unitary Character of Late Medieval Learning." The Cultural Context of Medieval

*Learning*. Edited by John E. Murdoch and Edith D. Sylla. Dordrecht: Reidel, 1975, pp. 271-348.

Argues that in addition to the virtual identity between science and philosophy, there was likewise a "weaker" unity between medieval philosophy and theology. Here, as in the case of the link between science and philosophy, the new analytical tools and languages of the fourteenth century assumed the same methodologically connecting function. See item 669.

670. Murdoch, John E. "The Development of a Critical Temper: New Approaches and Modes of Analysis in Fourteenth-Century Philosophy, Science, and Theology." *Medieval and Renaissance Studies* (Southeastern Institute of Medieval and Renaissance Studies), 7 (1978), 51-79.

Argues that fourteenth-century philosophical thought was distinguished by changes in methodology involving new "analytical languages," an increased tendency to treat problems in the context of propositional analysis, and the development of a critical and hesitant attitude which drew back from definitive conclusions when faced with alternative assumptions.

671. Murdoch, John E. "The Analytic Character of Late Medieval Learning: Natural Philosophy without Nature." *Approaches to Nature in the Middle Ages*. Edited by Lawrence D. Roberts. Binghamton, N.Y.: Center for Medieval and Early Renaissance Studies, 1982, pp. 171-213.

Argues that philosophy in the fourteenth century maintained a close relationship with theology. Philosophers utilized techniques of analysis which were applied to theological concerns. Argues likewise that it is possible to overemphasize the fourteenth century trend toward empiricism. Rather logical and mathematical methods, new "languages of analysis" were applied to problems *secundum imaginationem* in which experience and observation played little role. This is illustrated by examples of the medieval treatment of indivisibles. See item 625.

672. Randall, Jr., John Herman. *The School of Padua and the Emergence of Modern Science*. Padua: Antenore, 1962. Pp. 141.

Argues in the title essay that Aristotelians at the north Italian universities, particularly the Averroists at Padua during the first half of the fifteenth century, were the heirs of scientific methodology previously developed at Oxford and Paris in the thirteenth and fourteenth centuries. From about 1400 through the sixteenth century, several generations of Paduan scholars led Europe in discussions and elaborations of methodology, a development aided by the anticlerical nature of Italian institutions, the close connection between natural philosophy and medicine and the intellectual freedom offered by Venice. This essay was originally published in The Journal of the History of Ideas, 1 (1940), 177-206 without the Latin texts now included.

673. Sylla, Edith D. "The a posteriori Foundations of Natural Science: Some Medieval Commentaries on Aristotle's Physics, Book I, Chapters 1 and 2." Synthèse, 40 (1979), 147-187.

Argues from an examination of their Physics commentaries, that a group of schoolmen at Oxford had a view of physics as a discipline with its own foundations and methods not dependent on a priori principles.

674. Wallace, William A. The Scientific Methodology of Theodoric of Freiberg. Fribourg: The University Press, 1959. Pp. xvii + 395.

Presents Theodoric's methodological development as a "case study" illustrating the interplay between philosophy and science as seen in the writings of an Aristotelian scholastic who, as an experimental scientist, made an important contribution to an understanding of the rainbow. Argues that Theodoric's qualitative and causal approach to natural philosophy altered in the course of his optical works. In his De iride, he revealed a new methodology containing quantitative elements, and particularly a dependence on observation and experimental methods. It was this methodology which, while based on Aristotle, employed experimental procedures and enabled Theodoric to formulate the cause of the upper and lower rainbow.

675. Wallace, William A. Causality and Scientific Explanation. Vol. I: Medieval and Early Classical

Science. Ann Arbor: University of Michigan Press, 1972. Pp. xi + 288.

Argues, as a realist dissatisfied with positivism and its emphasis on logical rather than ontological explanation, that a historical treatment of the attitudes toward causality and scientific explanation in Greek and medieval thought can be instructive in the present. This volume traces the role of causal analysis and explanatory methodologies in the work of natural philosophers at Paris, Oxford and Padua, and maintains that early modern science demonstrates a methodological continuity with the medieval past.

676. Wallace, William A. "The Scientific Methodology of St. Albert the Great." Albertus Magnus, Doctor Universalis, 1280/1980. Edited by Gerbert Meyer and Albert Zimmermann. Mainz: Matthias-Grünewald-Verlag, 1980, pp. 385-407.

Characterizes Albert's methodology, basically Aristotelian, as revealed in his works on animate nature and on the physical world, in relation to both his own time and in a modern context.

## LOGIC: GENERAL

677. Boehner, Philotheus. Medieval Logic: An Outline of its Development from 1250 to c. 1400. Manchester: Manchester University Press, 1952. Pp. xvii + 130.

Outlines the sources of the logica antiqua, but focuses on the systems of logic developed by scholastics (the logica moderna) such as Peter of Spain, William of Ockham, Walter Burley, Albert of Saxony and John Buridan.

678. Kneale, William and Martha Kneale. The Development of Logic. Oxford: Clarendon Press, 1962. Pp. viii + 761.

A survey of logic from pre-Aristotelian developments to the twentieth century. Chapters II and III (pp. 23-176) on Aristotle's Organon and Megarian and Stoic logic are valuable for background. Chapter III (pp. 177-297) covers Roman and medieval logic.

679. Moody, Ernest A. "The Medieval Contribution to Logic." Studies in Medieval Philosophy, Science, and Logic: Collected Papers, 1933-1969. Berkeley: University of California Press, 1975, pp. 371-392.

Discusses the historical development of medieval logic and its position vis-à-vis Aristotelian and modern logic. During its creative period (logica moderna), medieval logic like natural philosophy, while developing new doctrines in order to confront difficulties in the Aristotelian heritage, remained within the framework of Aristotelian thought.

## LOGIC: SPECIFIC TOPICS

680. Abelard, Peter. Dialectica. Edited by L.M. de Rijk. Assen: van Gorcum, 1956. Pp. xcvii + 637.

681. Aegidius Romanus. Super libros Posteriorum analyticorum. Venice, 1488. Reprint. Frankfort: Minerva, 1967.

682. Aegidius Romanus. Expositio in artem veterem. Venice, 1507. Reprint. Frankfort: Minerva, 1968.

683. Aegidius Romanus. In libros priorum analyticorum expositio. Venice, 1516. Reprint. Frankfort: Minerva, 1968.

684. Albert of Saxony. Perutilis logica. Venice, 1522. Reprint. Hildesheim: Olms, 1974. Pp. 103.

685. Albert of Saxony. Sophismata. Paris, 1502. Reprint. Hildesheim: Olms, 1975.

686. Algazel (al-Ghazzālī). Logica et philosophia. Venice, 1506. Reprint with an introduction by Charles Lohr. Frankfort: Minerva, 1969. Pp. 148.

687. Ammonius. Commentaire sur le Peri hermeneias d'Aristote. Traduction de Guillaume de Moerbeke. Edited by G. Verbeke with a study of the utilization of the commentary in the work of St. Thomas. Paris: Béatrice-Nauwelaerts, 1961. Pp. cxx + 515.

An edition of the Moerbeke translation made in 1268, introduced by essays on Aquinas' use of the commentary, the doctrine of truth and contingency in Ammonius, Boethius and Aquinas, and Moerbeke's translation techniques.

688. Aquinas, Thomas. In Aristotelis libros Peri hermeneias et Posteriorum analyticorum expositio. Edited by R.M. Spiazzi. Turin: Marietti, 1955. Pp. xviii + 439.

689. Aquinas, Thomas. Exposition of the Posterior Analytics. Translated by Pierre Conway. Quebec: La libraire philosophique M. Doyon, 1956. Pp. xvi + 449.

690. Aquinas, Thomas. Commentary on the Posterior Analytics of Aristotle. Translated by F.R. Larcher with a preface by James A. Weisheipl. Albany, N.Y.: Magi, 1970. Pp. xi + 241.

691. Aristotle. Analytica posteriora. Translatio anonyma. Edited by Lorenzo Minio-Paluello. Aristoteles latinus, IV, 2. Bruges: Desclée de Brouwer, 1953. Pp. xiv + 111.

692. Aristotle. Analytica posteriora Gerardo Cremonensi interprete. Edited by Lorenzo Minio-Paluello. Aristoteles latinus, IV, 3. Bruges: Desclée de Brouwer, 1954. Pp. xxxv + 139.

693. Aristotle. Categoriae vel praedicamenta. Translatio Boethii-editio composita, translatio Guillelmi de Moerbeka, lemmata e Simplicii commentario decerpta, pseudo-Augustini paraphrasis Themistiana. Edited by Lorenzo Minio-Paluello. Aristoteles latinus, I, 1-5. Bruges: Desclée de Brouwer, 1961. Pp. xcvi + 257.

694. Aristotle. Analytica priora. Translatio Boethii (recensiones duae). Translatio anonyma. Pseudo-Philoponi aliorumque scholia. Specimina translationum recensiorum. Edited by Lorenzo Minio-Paluello. Aristoteles latinus, III, 1-4. Bruges: Desclée de Brouwer, 1962. Pp. xxxviii + 503.

695. Aristotle. De interpretatione vel Peri ermenias. Translatio Boethii, specimina translationum recentiorum. Edited by Lorenzo Minio-Paluello. Translatio Guillelmi de Moerbeka. Edited by Gerard

Verbeke, revised by Lorenzo Minio-Paluello. Aristoteles latinus, II, 1-2. Bruges: Desclée de Brouwer, 1965. Pp. xix + 128.

696. Aristotle. Categoriarum supplementa. Porphyrii Isagoge translatio Boethii et anonymi fragmentum vulgo vocatum Liber sex principiorum. Edited by Lorenzo Minio-Paluello and Bernard G. Dod. Aristoteles latinus, I, 6-7. Bruges: Desclée de Brouwer, 1966. Pp. xviii + 132.

697. Aristotle. Analytica posteriora. Translationes Iacobi, anonymi sivi "Ioannis," Gerardi, et recensio Guillelmi de Moerbeka. Edited by Lorenzo Minio-Paluello and Bernard G. Dod. Aristoteles latinus, IV, 1-4. Bruges: Desclée de Brouwer, 1968. Pp. xxxiii + 446.

698. Aristotle. Topica. Translatio Boethii, fragmentum recensionis alterius, et translatio anonyma. Edited by Lorenzo Minio-Paluello and Bernard G. Dod. Aristoteles latinus, V, 1-3. Brussels: Desclée de Brouwer, 1969. Pp. 375.

699. Aristotle. De sophisticis elenchis. Translatio Boethii, fragmenta translationis Iacobi et recensio Guillelmi de Moerbeke. Edited by Bernard G. Dod. Aristoteles latinus, VI, 1-3. Leiden: Brill, 1975. Pp. xlii + 152.

700. Averroes (Ibn Rushd). Middle Commentary on Porphyry's Isagoge, and on Aristotle's Categories. Translated by Herbert A. Davidson. Cambridge, Mass.: Mediaeval Academy of America, 1969. Pp. xxi + 126.

The Isagoge is translated from the Hebrew and Latin versions. the Categories from the original Arabic and the Hebrew and Latin versions.

701. Bacon, Roger. "The Unedited Part of Roger Bacon's Opus maius: De signis." Edited by K.M. Fredborg, Lauge Nielsen and Jan Pinborg. Traditio, 34 (1978), 75-136.

Provides the text of what is likely a missing part of the Opus maius. See item 745.

702. Billingham, Richard. "Lo Speculum puerorum sive Terminus est in quem di Riccardo Billingham."

Edited by Alfonso Maierù. Studi medievali, 3 (1969), 297-397. See item 750.

703. Boethius. De topicis differentiis. Translated with notes and essays on the text by Eleanore Stump. Ithaca: Cornell University Press, 1978. Pp. 287.

704. Bottin, Francesco. "Un testo fondamentale nell'ambito della nuova fisica di Oxford: I Sophismata di Richard Kilmington." Antiqui und Moderni: Traditionsbewusstsein und Fortschrittsbewusstsein im späten Mittelalter. Edited by Albert Zimmermann. Berlin: de Gruyter, 1974, pp. 201-205.

Draws attention to the surviving MSS of the Sophismata and to Kilmington's place among the mathematicians and logicians of fourteenth-century Oxford.

705. Brown, Stephen F. "Walter Burleigh's Treatise De suppositionibus and its Influence on William of Ockham." Franciscan Studies, 32 (1972), 15-64.

Presents an edition of Burley's De suppositionibus, an early work probably written around 1302, and discusses the relationship between Burley and Ockham on the subject of supposition. Points out that Ockham used this early Burley work as one of his sources in the Summa logicae, and that where he opposes Burley's opinions, Ockham is arguing against traditional and accepted positions, not just Burley. Agrees that older views of Burley as a mediocre logician are not justified; he was a relatively original author, but fundamentally traditional.

706. Brown, Stephen F. "A Modern Prologue to Ockham's Natural Philosophy." Sprache und Erkenntnis im Mittelalter: Akten des VI. internationalen Kongresses für mittelalterliche Philosophie, I. Berlin: de Gruyter, 1981, pp. 107-129.

Discusses Ockham's theory of language as applied to his natural philosophy. In his theory of supposition, his treatment of the Aristotelian categories, in his adamant denial of any reification, Ockham created an interpretation of Aristotelian natural philosophy to which his logical work provides a guide.

707. Buridan, John. "Giovanni Buridano: Tractatus de suppositionibus." Edited by Maria Elena Reina. Rivista critica di storia della filosofia, 12 (1957), 175-208, 323-352.

708. Buridan, John. Compendium totius logicae. Venice, 1499. Reprint. Frankfort: Minerva, 1965.

709. Buridan, John. Sophisms on Meaning and Truth. Translated by Theodore K. Scott. New York: Appleton-Century-Crofts, 1966. Pp. xv + 223.

A translation of Buridan's Sophismata. See item 711.

710. Buridan, John. Tractatus de consequentiis. Edited by Hubert Hubein. Louvain: Publications universitaires, 1976. Pp. 138.

711. Buridan, John. Sophismata. Edited by Theodore K. Scott. Stuttgart-Bad Cannstatt: Frommann-Holzbook, 1977. Pp. 168.

An edition of the text.

712. Burley, Walter. De puritate artis logicae: Tractatus longior. With a Revised Edition of the Tractatus brevior. Edited by Philotheus Boehner. St. Bonaventure, N.Y.: Franciscan Institute, 1955. Pp. xvi + 264.

713. Burley, Walter. "Burleigh: On Conditional Hypothetical Syllogisms." Translated by Ivan Boh. Franciscan Studies, 23 (1963), 4-67.

A translation of Tract II, Part 1, Chapters 1-3 of Burley's De puritate artis logicae--Tractatus longior from the edition by Philotheus Boehner. See item 712.

714. Burley, Walter. "De primo et ultimo instanti." Edited by Herman and Charlotte Shapiro. Archiv für Geschichte der Philosophie, 47 (1965), 157-173.

Argues that a necessary revaluation of Burley as an independent thinker requires further study of his writings and provides an edition of this treatise based on three fourteenth-century MSS.

715. Burley, Walter. Super artem veterem. Venice, 1497. Reprint. Frankfort: Minerva, 1967.

716. Grabmann, Martin. "Kommentare zur aristotelischen Logik aus dem 12. und 13. Jahrhundert im Ms. lat. fol. 624 der Preussischen Staatsbibliothek in Berlin: Ein Beitrag zur Abaelardforschung." Sitzungsberichte der Preussischen Akademie der Wissenschaften, phil.-hist. Klasse, 18 (1938), 185-210.

Analyzes the content of the MS originally from the abbey of St. Victor in Paris, containing chiefly commentaries on the old logic, several of which reflect the opinions of Abelard (one is by Abelard).

717. Grabmann, Martin. Die Sophismataliteratur des 12. und 13. Jahrhunderts mit Textausgabe eines Sophisma des Boetius von Dacien. Beiträge zur Geschichte des Philosophie und Theologie des Mittelalters, 36, 1. Münster: Aschendorff, 1940. Pp. viii + 98.

Traces the development of the sophismata-literature during the twelfth to the beginning of the fourteenth century, with representative examples, including the text of Boetius' sophism: Omnis homo de necessitate est animal.

718. Hendley, Brian. "John of Salisbury and the Problem of Universals." Journal of the History of Philosophy, 8 (1970), 289-302.

Maintains that John of Salisbury made an original contribution to the problem of universals differing both from that of his contemporaries and that of Aristotle.

719. Heytesbury, William. William Heytesbury on "Insoluble" Sentences. Translated by Paul V. Spade. Toronto: Pontifical Institute of Mediaeval Studies, 1979. Pp. 111.

A translation of Chapter 1 of Heytesbury's Regule solvendi sophismata from the Venice, 1494 edition.

720. Kilmington, Richard. "L'Opinio de insolubilibus de Richard Kilmington." Edited by Francesco Bottin. Rivista critica di storia della filosofia, 28 (1973), 408-421.

An edition of the last sophism of Kilmington: _A est scitum a te_.

721. Kretzmann, Norman. "Incipit/Desinit." _Motion and Time, Space and Matter: Interrelations in the History of Philosophy and Science_. Edited by Peter K. Machamer and Robert G. Turnbull. Columbus: Ohio State University Press, 1976, pp. 101-136.

Argues that the concern of logicians with _incipit/desinit_ problems arose in twelfth-century treatments of fallacies stemming from _De sophisticis elenchis_ which predated discussions introduced by the _Physics_.

722. Lambert of Auxerre. _Logica (Summa Lamberti)_. Edited by Franco Alessio. Florence: La nuova Italia, 1971.

723. de Libera, A. "Le traité _De appellatione_ de Lambert de Lagny (Lambert d'Auxerre)." _Archives d'histoire doctrinale et littéraire du moyen âge_, 48 (1981), 227-285.

Discusses the sources and content of the eighth tract of the _Summa Lamberti_ with the text. See item 722.

724. Lohr, Charles H. "_Logica Algazelis_: Introduction and Critical Text." _Traditio_, 21 (1956), 223-290.

Discusses the background of al-Ghazzāli's _The Meanings of the Philosophers_, translated in the second half of the twelfth century as the _Liber Algazelis de summa theoricae philosophiae_. Several parts of the work circulated independently such as the _Tractatus de logica_ edited here.

725. Maierù, Alfonso. "Il _Tractatus de sensu composito et diviso_ di Guglielmo Heytesbury." _Rivista critica di storia della filosofia_, 21 (1966), 243-263.

Describes Heytesbury's treatment of a logical doctrine which exercised much influence on fifteenth-century Italian commentators. The analysis of propositions in a composite or divisive sense stems from the _De sophisticis elenchis_.

726. Marsilius of Inghen. _Commentum in primum et quartum tractatus Petri Hispani_. Reprint. Frankfort: Minerva, 1967.

727. Moody, Ernest A. *The Logic of William of Ockham.* New York: Sheed & Ward, 1935. Pp. xiv + 322.

Provides an analysis of Ockham's nominalistic interpretation of the *Organon* based primarily on his *Summa totius logicae* and the *Expositio aurea super artem veterem*.

728. Moody, Ernest A. *Truth and Consequence in Mediaeval Logic*. Amsterdam: North-Holland, 1953. Pp. viii + 113.

Investigates two branches of medieval logical analysis, the theory of supposition and the theory of consequences or of truth-conditions or of inference-conditions or deduction. These medieval theories, developed within the context of the formal logic of the period, is studied primarily in three fourteenth-century works: the *Summa logicae* of Albert of Saxony, and the *Sophismata* and *Consequentiae* of Jean Buridan.

729. Ockham, William of. *Summa logicae*: *Pars prima*. *Pars secunda et tertiae prima*. 2 vols. Edited by Philotheus Boehner. St. Bonaventure, N.Y.: Franciscan Institute, 1957, 1962. Pp. xiv + 461.

730. Ockham, William of. *Expositio aurea et admodum utilis super Artem veterem ... cum questionibus Alberti parvi de Saxonia*. Bologna, 1496. Reprint. Farnborough: Gregg, 1964.

731. Ockham, William of. *Expositionis in libros artis logicae prooemium et Expositio in librum Porphyrii De predicalibus*. Edited by Ernest A. Moody, St. Bonaventure, N.Y.: Franciscan Institute, 1965. Pp. 147.

732. Ockham, William of. *Ockham's Theory of Terms*: *Part I of the Summa logicae*. Translated and introduced by Michael J. Loux. Notre Dame: University of Notre Dame Press, 1974. Pp. xiii + 221.

Introduces the translation with essays on the ontology of Ockham and Ockham on generality. See item 729.

733. Ockham, William of. *Venerabilis inceptoris Guillelmi de Ockham Summa logicae*. Edited by Philotheus

Boehner, Gedeon Gál and Stephen Brown. St. Bonaventure, N.Y.: St. Bonaventure University, 1974. Pp. 73, 899.

734. Ockham, William of. Expositio in librum praedicamentorum Aristotelis. Edited by G. Gál. St. Bonaventure, N.Y.: St. Bonaventure University, 1978.

735. Ockham, William of. Ockham's Theory of Propositions. Part II of the Summa logicae. Translated by Alfred J. Freddoso and Henry Schuurman with an introduction by Alfred J. Freddoso. Notre Dame: University of Notre Dame Press, 1980. Pp. vii + 212.

The translation is introduced by an essay on Ockham's theory of truth conditions. See item 729.

736. O'Donnell, J. Reginald. "Themistius' Paraphrasis of the Posterior Analytics in Gerard of Cremona's Translation." Mediaeval Studies, 20 (1958), 239-315.

An edition of the text.

737. Paul of Pergula. Logica and Tractatus de sensu composito et diviso of Paul of Pergula. Edited by Sister Mary Anthony Brown. St. Bonaventure, N.Y.: Franciscan Institute, 1961. Pp. xiv + 162.

Editions of works by this fifteenth-century Italian nominalist.

738. Paul of Pergula. "On Supposition and Consequences." Translated by Ivan Boh. Franciscan Studies, 25 (1965), 30-89.

A translation of three tracts, "On Suppositions," "On Hypothetical Propositions," and "On Consequences," from the edition of Paul's logic by Sister Mary Anthony Brown. See item 737.

739. Paul of Venice. Pauli Veneti Logica magna, secunda pars. Tractatus de veritate et falsitate propositionis et Tractatus de significato propositionis. Edited with notes on the sources by Francesco del Punta. Translated with notes by Marilyn McCord Adams. Oxford: Oxford University Press, 1978.

740. Peter of Spain. The Summulae logicales of Peter of Spain. Edited by Joseph P. Mullally. Notre Dame: University of Notre Dame Press, 1945. Pp. civ + 172.

Discusses the content of Tracts I-VI, and provides an edition and translation of Tract VII (De proprietatibus terminorum). See item 742.

741. Peter of Spain. Tractatus syncategorematum and Selected Anonymous Treatises. Translated by Joseph P. Mullally, with an introduction by Joseph P. Mullally and Roland Honde. Milwaukee: Marquette University Press, 1964. Pp. ix + 156.

Translations of Peter's treatise on syncategorematic terms and of his treatises on obligations, insolubles and consequences. Taken from the Summulae logicales. See item 742.

742. Peter of Spain. Peter of Spain Tractatus Called Afterward Summule logicales. Edited by L.M. de Rijk. Assen: van Gorcum, 1972. Pp. xxix + 313.

743. Pinborg, Jan. "The English Contribution to Logic before Ockham." Synthèse, 40 (1979), 19-42.

Seeks the factors which may explain the flowering of the application of terminist logic to numerous areas of natural philosophy which began in the fourteenth century, especially at Oxford. Surveys the logical contributions of English masters before Ockham and suggests that at Oxford the terminist tradition which began in the twelfth century was maintained without the break which occurred at Paris.

744. Pinborg, Jan. "A Logical Treatise Ascribed to Bradwardine." Studi sul XIV secolo in memoria Anneliese Maier. Edited by Alfonso Maierù and Agostino Paravicini Bagliani. Rome: Edizioni di storia e letteratura, 1981, pp. 27-55.

Provides an edition of a logical text--Opus artis logicae--which may be but likely is not by Bradwardine, consisting of three chapters on respectively syncategorematic terms, their use in sophisms, and on distinctions de composito et diviso.

745. Pinborg, Jan. "Roger Bacon on Signs: A Newly Discovered Part of the *Opus maius*." *Sprache und Erkenntnis im Mittelalter: Akten des VI. internationalen Kongresses für mittelalterliche Philosophie*, I. Berlin: de Gruyter, 1981, pp. 403-412.

Discusses Bacon's treatment of the concept of signs and their function in language. Expresses a high valuation of Bacon's semantic theories.

746. Reina, Maria Elena. "Il problema del linguaggio in Buridano." *Rivista critica di storia della filosofia*, 14 (1959), 367-417; 15 (1960), 141-165, 238-264.

Analyzes Buridan's semantic-linguistic theories so as to cast light on his position vis-à-vis the Paris arts faculty statute of 1340. Concludes that Buridan signed the statute with no violence to his convictions, although his opinions were in some respects closer to those of Ockham than to the official position.

747. de Rijk, L.M. "Some New Evidence on Twelfth-Century Logic: Alberic and the School of Mont Ste. Geneviève (Montani)." *Vivarium*, 4 (1966), 1-57.

Presents new MS information on the teaching of logic at the school of Ste. Geneviève, especially that of a Master Alberic of Paris.

748. de Rijk, L.M. "Some Notes on the Mediaeval Tract *De insolubilibus*, with an Edition of a Tract Dating from the End of the Twelfth Century." *Vivarium*, 4 (1966), 83-115.

Traces the medieval sources for puzzles such as the liar paradox which arise from the self-reflexive character of statements. Complex types flourished from the thirteenth century in tracts *De insolubilibus*. Gives the text of what is likely the oldest such work and compares it with others in the genre.

749. de Rijk, L.M. *Logica modernorum: A Contribution to the History of Early Terminist Logic*. 2 vols. Assen: van Gorcum, 1962, 1967.

Argues that the *logica moderna* had roots in grammatical studies in centers at Paris, Oxford and perhaps northern Italy before the thirteenth century. Vol. I deals with one of these roots--twelfth-century theories of fallacy. Vol. II (Part 1) traces a second source in the development of linguistic-semantic theories basic to the theory of supposition. Traces the background of this theory as well as the origins of treatises on syncategorematic terms and other features of the logica moderna. Part 2 contains editions of relevant texts.

750. de Rijk, L.M. "Another Speculum puerorum Attributed to Richard Billingham: Introduction and Text." Medioevo, 1 (1975), 203-235.

A treatise on the truth and falsity of propositions similar to that edited by Maierù, also called the Speculum puerorum and likewise ascribed to Billingham. Despite their similarities, the two works do not represent two MS traditions of the same tract. See item 702.

751. Simplicius. Commentaire sur les Catégories d'Aristote. Traduction de Guillaume de Moerbeke. 2 vols. Edited by A. Pattin and W. Storyven. Louvain: Publications universitaires, 1971, 1975. Pp. liv + 765.

752. Spade, Paul V. "An Anonymous Tract on Insolubilia from MS Vat. lat. 674: An Edition and Analysis of the Text." Vivarium, 9 (1971), 1-18.

The text of a tract written in 1368.

753. Spade, Paul V. "The Treatises On Modal Propositions and On Hypothetical Propositions by Richard Lavenham." Mediaeval Studies, 35 (1973), 49-59.

Provides editions of two tracts by Lavenham (d. late fourteenth century) and analyzes the contents of a British Library MS containing Lavenham treatises, chiefly on logic, including these two.

754. Spade, Paul V. "Ockham's Distinctions between Absolute and Connotative Terms." Vivarium, 13 (1975), 55-76.

Treats Ockham's distinction between absolute and connotative terms, particularly in his Summa logicae, clarifying ideas connected with the distinction and maintaining that in the aspects considered here, Ockham's connotation theory demands no "new primitive notions" apart from those needed for absolute terms.

755. Spade, Paul V. "Some Epistemological Implications of the Burley-Ockham Dispute." Franciscan Studies, 35 (1975), 212-222.

Argues that the differences between Burley and Ockham over supposition and signification were based on more than the metaphysical disagreement usually the focus of discussion. The difference in their views is likewise an epistemological one stemming from their diverse interpretations of signification which determined the kind of supposition a term had when it stood for what it signified.

756. Spade, Paul V. "Roger Swyneshed's Obligationes: Edition and Comments." Archives d'histoire doctrinale et littéraire du moyen âge, 44 (1977), 243-285.

Introduces the edition by remarking that the obligationes-literature has not been fully studied; the exact purpose of this kind of logical confrontation is unknown.

757. Spade, Paul V. "Roger Syneshed's Insolubilia: Edition and Comments." Archives d'histoire doctrinale et littéraire du moyen âge, 46 (1979), 177-220.

The text of a treatise on sophisms similar to the liar paradox.

758. Sylla, Edith D. "William Heytesbury on the Sophism 'Infinita sunt finita'." Sprache und Erkenntnis im Mittelalter: Akten des VI. internationalen Kongresses für mittelalterliche Philosophie, II. Berlin: de Gruyter, 1981, pp. 628-636.

Analyzes Heytesbury's discussion of this sophism treated within the theory of supposition. While problems concerning the infinite arose in the context of natural philosophy, medieval logicians treated them as a part of logic and the theory of language.

759. Synan, Edward A. "The Insolubilia of Roger Nottingham, O.F.M." Mediaeval Studies, 26 (1964), 257-270.

Gives the text of a tract on insolubles written when Nottingham was a bachelor at Oxford.

760. Webering, Damascene. Theory of Demonstration According to William of Ockham. St. Bonaventure, N.Y.: Franciscan Institute, 1953. Pp. xii + 186.

Analyses Ockham's views on the nature and object of demonstration, using material taken from the Summa totius logicae and the prologue to the commentary on the Sentences.

761. William of Sherwood. Die Introductiones in logicam. Edited by Martin Grabmann. Sitzungsberichte der Bayerischen Akademie der Wissenschaften, 10. Munich, 1937. Pp. 106.

An edition of the text.

762. William of Sherwood. Treatise on Syncategorematic Words. Translated with an introduction and notes by Norman Kretzmann. Minneapolis: University of Minnesota Press, 1968. Pp. 173.

763. William of Sherwood. Introduction to Logic. Translated by Norman Kretzmann. Westport, Conn.: Greenwood Press, 1975. Pp. xiii + 187.

See item 761.

764. Yokoyama, Teksuo. "Simon of Faversham's Sophisma: Universale est intentio." Mediaeval Studies, 31 (1969), 1-14.

A sophism by Faversham who taught at Oxford around the beginning of the fourteenth century.

# MATHEMATICS

## GENERAL

765. von Braunmühl, A. *Vorlesungen über Geschichte der Trigonometrie*. 2 pts. Leipzig: Teubner, 1900, 1903.

   A survey of the history of trigonometry from its earliest beginnings including Islamic developments. Chapters 6 and 7 cover western Europe in the Middle Ages.

766. Busard, H.L.L. *Quelques subjets de l'histoire des mathématiques au moyen âge*. Paris: Palais de la Découverte. 1968, Pp. 32.

   Surveys mathematical thought from the tenth to the fifteenth centuries as divided into four periods: (1) The mathematical Renaissance of the late tenth century introduced by Gerbert; (2) the translation of Arabic materials and the change of emphasis from practical to theoretical mathematics; (3) a period of independent mathematical work; (4) the application of mathematics to physics and the treatment of infinite series at Oxford and Paris.

767. Folkerts, Menso. "Regiomontanus als Mathematiker." *Centaurus*, 21 (1977), 214-245.

   Evaluates Regiomontanus as an extremely talented and capable mathematician. His mathematical activity was cut short by his premature death and the subsequent dispersal of his works.

768. Gerbert (Pope Sylvester II). Gerberti ... opera mathematica (972-1003). Edited by Nicolaus Bubnov. Berlin, 1899. Reprint. Hildesheim: Olms, 1963. Pp. cxix + 620.

Editions of the genuine and doubtful works of Gerbert with appendixes which contain related materials.

769. Juschkewitsch, A.P. Geschichte der Mathematik im Mittelalter. Translated from Russian by Viktor Ziegler. Leipzig: Teubner, 1964. Pp. viii + 454.

The Russian original appeared in 1961. A survey of the chief characteristics of mathematics in China, India and Islam. The final chapter surveys western Europe from Bede and Alcuin to the sixteenth century.

770. Leonardo of Pisa (Fibonacci). Scritti di Leonardo Pisano. 2 vols. Edited by Baldassarre Boncompagni. Rome: Tip. della scienze matematische, 1857-1862. Pp. 464, 288.

## ARITHMETIC, ALGEBRA, INFINITE SERIES, NUMBER THEORY

771. Alexander of Villedei. Carmen de algorismo. Rara mathematica. 2d ed. Edited by J.O. Halliwell. London, 1841. Reprint. Hildesheim: Olms, 1977, pp. 73-83.

See item 817.

772. Allard, André. "A propos d'un algorisme latin de Frankenthal: Une méthode de recherche." Janus, 65 (1978), 119-141.

Analyzes the text of an ars algorismi, originally from a monastery in Frankenthal, by comparing selected characteristics of the arithmetical operations in the treatise with those of ten algorisms of the twelfth and thirteenth centuries, including the works of Sacrobosco and Alexander of Villedieu.

773. Arrighi, Gino, ed. Libro d'abaco dal codice 1754 (sec. XIV) della Biblioteca Statale di Lucca. Lucca: Cassa di risparmio di Lucca, 1973. Pp. 205.

An edition of an Italian work on the abacus containing arithmetical problems which were germane to the daily experience of Lucca merchants.

774. Boethius. De institutione arithmetica, libri duo. De institutione musica, libri quinque. Edited by G. Friedlein. Leipzig: Teubner, 1867. Pp. viii + 492.

775. Busard, H.L.L. "Über unendlichen Reihen im Mittelalter." L'enseignement mathématique, sér, 2, 8, (1962), 281-290.

Discusses the geometric and algebraic methods of summing infinite series found in the medieval literature.

776. Busard, H.L.L. "Unendliche Reihen in A est unum calidum." Archive for History of Exact Sciences, 2 (1965), 387-397.

Discusses and gives the text of a short treatment of infinite series attached to a Bibliothèque Nationale MS of A est unum calidum.

777. Busard, H.L.L. "L'algèbre au moyen âge: Le Liber mensurationum d'Abū Bekr." Journal des savants, (1968), 65-124.

An edition of Abū Bekr's treatise, a work on measurement which supplied a theoretical base for practical geometry. The work applies algebra to geometry, using equations of the first and second degree as in the first part of al-Khwārizmī's work on algebra, and likewise, employs algebraic methods of Babylonian origin. The application of algebra to problems of measurement entered the Latin West with al-Khwārizmī, Savasorda and Abū Bekr.

778. Busard, H.L.L. "Die Arithmetica speculativa des Johannes de Muris." Scientiarum historia, 13 (1971), 103-132.

An edition of a popular text on the theory of numbers and algebra written in 1324.

779. Charmasson, Thérèse. "L'arithmétique de Roland l'Ecrivain et le Quadripartitum numerorum de Jean de Murs." Revue d'histoire des science, 31 (1978), 173-176.

Maintains that the arithmetic work by Roland, physician to the Duke of Bedford between 1225-1235, is almost completely identical with the Quadripartitum of John de Muris.

780. Curtze, Maximilian. "Die Ausgabe von Jordanus De numeris datis durch Professor P. Treutlein in Karlsruhe." Leopoldina, amtlichen Organ der kaiserlichen Leopoldina-Carolinischen Deutschen Akademie der Naturforscher, 18 (1882), 26-31.

Contains corrections to Treutlein's edition. See item 802.

781. Curtze, Maximilian. "Commentar zu dem Tractatus de numeris datis." Zeitschrift für Mathematik und Physik, hist.-lit. Abtheilung, 36 (1891), 1-138.

A more complete text of the work than in Treutlein's edition. See item 802.

782. Curtze, Maximilian. "Über eine Algorismus-Schrift des XII. Jahrhunderts." Abhandlungen zur Geschichte der Mathematik, 8 (1898), 1-27.

Describes the content of an algorism in three sections (whole numbers, fractions, extraction of square roots) which serves as part of the introduction to a Tractatus astronomicus anonymi in a Munich MS. Provides the Latin text of the sections.

783. van Egmond, Warren. "The Earliest Vernacular Treatment of Algebra: The Libro di ragioni of Paolo Gerardi (1328)." Physis, 20 (1978), 155-189.

Provides the text with translation of the algebraic section of a work by Gerardi of Florence, written at Montpellier in 1328. It includes 9 kinds of cubic equations, the earliest known treatment of cubics in western Europe.

784. Endrei, Walter. "De l'abaque aux chiffres arabes, leur lutte en Europe." Studi in memoria di Federigo Melis, I. Naples: Giannini, 1978, pp. 279-300.

Traces the introduction of Arabic numerals into the Latin West and their relationship with the abacus. Discusses the difficulties in the acceptance of these numerals due to psychological pressures, the confusion of non-decimal monetary systems and weights. The triumph of Arabic numerals in Europe came with the evolution of teaching methods of calculation in influential manuals of arithmetic.

785. Eneström, Gustav. "Über die _Demonstratio Jordani de algorismo_." _Biblioteca mathematica_, ser. 3, 7 (1906-07), 24-37.

Provides the text of the definitions but not the proofs.

786. Eneström, Gustav. "Über eine dem Jordanus Nemorarius zugeschriebene kurze Algorismusschrift." _Biblioteca mathematica_, ser. 3, 8 (1907-08), 135-153.

The text with analysis of the introduction and the propositions with a few proofs of the treatise beginning: _Communis et consuetus rerum...._

787. Eneström, Gustav. "Der _Algorismus de integris_ des Meisters Gernardus." _Biblioteca mathematica_, ser. 3, 13 (1912-13), 289-332.

Text with commentary of the first part of the _Algorism_ of "Master Gernardus" with a comparison to similar material by Jordanus de Nemore. See item 788.

788. Eneström, Gustav. "Der _Algorismus de minutiis_ des Meisters Gernardus." _Biblioteca mathematica_, ser. 3, 14 (1913-14), 99-149.

Text and commentary of the second part of the _Algorism_ of "Master Gernardus" with notes on its relation to the work of Jordanus de Nemore. See item 787.

789. Eneström, Gustav. "Das Bruchrechnen des Jordanus Nemorarius." _Biblioteca mathematica_, ser. 3, 14 (1913-14), 42-44, 48-53.

The text of the introductions and propositions of the _Tractatus minutiarum_ and _Demonstratio de minutiis_ with proof of only two propositions.

790. Evans, Gillian R. "_Duc oculum_: Aids to Understanding in Some Mediaeval Treatises on the Abacus." _Centaurus_, 19 (1975), 252-263.

Points out that writers of treatises on the abacus, wishing to teach principles as well as practical applications of arithmetic, often relied on images and analogies, or visual aids, such as diagrams, so as to make difficult and tedious material more palatable.

791. Evans, Gillian R. "The _Rithomachia_: A Mediaeval Mathematical Teaching Aid?" _Janus_, 63 (1976), 257-273.

Argues that the _Rithomachia_, a kind of medieval mathematical chess, was used as an aid in the teaching of arithmetic during the twelfth century. It enabled students to master the relationships between numbers and to develop the skill to calculate readily. Early treatises emphasize its pedagogical use; the meager explanation of the rules they contain imply that the game was taught with the help of an experienced master. After the twelfth century, the game was played for pleasure by many; modified, with more detailed instructions, it remained popular in the sixteenth century.

792. Evans, Gillian R. "_Difficillima et ardua_: Theory and Practice in Treatises on the Abacus." _Journal of Medieval History_, 3 (1977), 21-38.

Argues that authors of abacus treatises emphasized more than the mechanics of calculation. Their works contain material on the _quadrivium_, especially geometry, treat the theory of number drawn from Boethius and reflect a growing interest in dialectic. Although such topics enlarged the scope of abacus treatises, these works offered no real possibilities for further development.

793. Evans, Gillian R. "From Abacus to Algorism: Theory and Practice in Medieval Arithmetic." _British Journal for the History of Science_, 10 (1977), 114-131.

Treats the change in elementary works on arithmetic during the thirteenth century from those on the abacus to treatises employing the method of calculation known as algorism.

794. Evans, Gillian R. "A Commentary on Boethius' Arithmetica of the Twelfth or Thirteenth Century." Annals of Science, 35 (1978), 131-141.

Maintains that an increasing interest in the Arithmetica of Boethius in the schools of northern France such as Paris, Laon and Chartres resulted in commentaries, one of which, in the Bayerische Staatsbibliothek, appears to be of the school of Hugh of St. Victor. This Victorine commentary omits the mathematical content of the Boethian treatise and focuses on a philosophical treatment of the study of number thus reflecting the interests of the twelfth-century schools.

795. Evans, Gillian R. "Introductions to Boethius' Arithmetica of the Tenth to the Fourteenth Century." History of Science, 16 (1978), 22-41.

Discusses the accessus genre as applied to arithmetical texts and analyzes several. Concludes that these introductions, which drew from the other artes, and which were designed to acquaint the young student with the author and subject he was about to study, could not have been helpful in the teaching of arithmetic.

796. Evans, Gillian R. "The Saltus Gerberti: The Problem of the 'Leap'." Janus, 67 (1980), 261-268.

Analyzes Gerbert's treatment of proportions and inequalities in his commentary on chapter one of Book II of Boethius' Arithmetica. The name, Gerbert's jump or leap, was attached to the passage to describe Gerbert's technique of reversing the order of the terms in a series so that the numbers could be considered as leaping about to face the other way.

797. Folkerts, Menso. "Mathematische Aufgabensammlungen aus dem ausgehenden Mittelalter: Ein Beitrag zur Klostermathematik des 14. und 15. Jahrhunderts." Sudhoffs Archiv: Zeitschrift für Wissenschaftsgeschichte, 55 (1971), 58-75.

Discusses the contents of 30 anonymous collections containing mathematical puzzles; only in rare cases is the method of solution given. These treatises contain standard problems; some important examples are given here.

798. Folkerts, Menso. "Pseudo-Beda: De arithmeticis propositionibus: Eine mathematische Schrift aus der Karolingerzeit." Sudhoffs Archiv: Zeitschrift für Wissenschaftsgeschichte, 56 (1972), 22-43.

An edition of a treatise in four parts. The first three sections provide various methods for guessing solutions to problems; the first appearance of this sort of recreational mathematics in the West. Part four has rules for the addition of positive and negative numbers unknown in the West before the fifteenth century.

799. Frajese, Attilio. "L'algebra dagli Arabi all'occidente." Oriente e occidente nel medioevo: Filosofia e scienze. Rome: Accademia nazionale dei Lincei, 1971, pp. 713-725.

Traces the development and transmission of algebra, limited to equations of the second degree, from Babylonia to Greece, from Greece to the Arabs, and then to the European West.

800. Glushkov, Stanislaw. "On Approximation Methods of Leonardo Fibonacci." Historia mathematica, 3 (1976), 291-296.

Argues that the tradition which considered approximation methods unsuitable for inclusion in mathematical treatises may be the reason why Leonardo omits the method that enabled him to calculate the root of a cubic equation with great accuracy. Argues that Leonardo calculated the approximate value by using the rule of two false positions in a step-by-step fashion.

801. Hughes, Barnabas B. "De regulis generalibus: a 13th-century English Mathematical Tract on Problem Solving." Viator, 11 (1980), 209-224.

An edition of a De regulis generalibus algorismi ad solvendum omnes questiones propositas, a short work of

English origin on solving problems with general rules and examples, probably designed for baccalaureates with some training in algorism.

801a. l'Huillier, Ghislaine. "Regiomontanus et le Quadripartitum numerorum de Jean de Murs." Revue d'histoire des sciences, 33 (1980), 193-214.

Discusses some of the marginal notes Regiomontanus provided on his MS copy of John of Murs' Quadripartitum (which the Viennese mathematician and astronomer had intended to publish at Nuremberg), pointing out Regiomontanus' preference for geometric rather than algebraic methods.

802. Jordanus de Nemore. "Der Traktat des Jordanus Nemorarius De numeris datis." Edited by P. Treutlein. Zeitschrift für Mathematik und Physik. Supplementheft, hist.-lit. Abteilung, 24 (1879), 125-166.

Latin text of the treatise. See items 780, 781.

803. Karpinski, Louis C. "The Quadripartitum numerorum of John of Meurs." Biblioteca mathematica, ser. 3, 13 (1912-13), 99-114.

Describes the general content and structure of Johannes de Muris mathematical treatise, pointing out its debt to the algebra of al-Khwārizmī and its similarities to the work of Jordanus de Nemore.

804. Karpinsky, Louis C. "Robert of Chester's Latin Translation of the Algebra of Al-Khowarizmi." Contributions to the History of Science. Louis C. Karpinski and John G. Winter. Ann Arbor: University of Michigan, 1930, pp. 1-164. Reprint. New York: Johnson Reprint, 1972.

Text and translation of the Algebra.

805. al-Khwārizmī. Mohammed ibn Musa Alchwarizmi's Algorismus: Das früheste Lehrbuch zum Rechnen mit indischen Ziffern. Edited by Kurt Vogel. Milliaria, 3. Aalen: Zeller, 1963. Pp. 49.

A facsimile and transcription of al-Khwārizmī's Arithmetic from the unique Latin MS (Cambridge University Library MS Ii. 6. 5) with commentary.

806. Lemay, Richard. "The Hispanic Origin of our Present Numeral Forms." Viator, 8 (1977), 435-462.

Argues from study of Latin and Arabic MSS of Spanish provenance on astronomy, astrology and mathematics that our number symbols are derived from Spanish use of Roman numerals which combined with Hindu numerals that had come into Spain. This created a set of Arabic numerals which were transmitted to the rest of Europe.

807. Leonardo of Pisa (Fibonacci). Leonard de Pise: De livre des nombres carres. Edited by Paul ver Eecke. Bruges: Desclée de Brouwer, 1952.

French translation of the Liber quadratorum.

808. Levey, Martin. The Algebra of Abū Kāmil ... in a Commentary by Mordecai Finzi. Madison: University of Wisconsin Press, 1966. Pp. xi + 226.

An edition of the Hebrew text with English translation, translated from Arabic by Finzi. Abū Kāmil's treatise was also translated into Latin and, with the work of al-Khwārizmī, provided the West with its chief Arabic algebraic sources.

809. Lubar, Steven. "The Algebra of Jordanus of Nemore: The Arabic Influence." Synthesis (Cambridge), 3 (1977), 33-43.

Compares the De numeris datis of Jordanus with the algebraic treatises of al-Khwārizmī and abū Kāmil, both available in Latin from the twelfth century. Concludes that Jordanus made no use of these Arabic algebraic works.

810. Masi, Michael. "The Influence of Boethius' De arithmetica on Late Medieval Mathematics." Boethius and the Liberal Arts. Edited by Michael Masi. Berne: Lang, 1981, pp. 81-95.

Argues that the Arithmetica continued to be an important text during the fourteenth and fifteenth centuries, both directly and in adaptations. Directly employed by John de Muris in the fourteenth century, its influence is likewise attested in many mathematical works in late medieval MS collections. The work also provided a starting point for fourteenth-century mathematicians such as Bradwardine and Albert of Saxony.

811. Nagl, Alfred. "Das Quadripartitum numerorum des Johannes de Muris und das prakische Rechnen im vierzehnten Jahrhundert." Abhandlungen zur Geschichte der Mathematik, 5 (1890), 137-146.

Discusses, within the context of the slow adoption of Hindu-Arabic numerals and the continued use of the abacus in practical reckoning, the sections of the Quadripartitum (Chapters 11 and 14 of Book II) which deal respectively with the easy multiplication of whole numbers and the abacus. Latin text of both chapters.

812. Nicomachus of Gerasa. Introduction to Arithmetic. Translated by Martin Luther d'Ooge. University of Michigan Studies, Humanistic Series, 16. New York: Macmillan, 1926. Pp. ix + 318.

Also contains studies in Greek arithmetic by Frank E. Robbins and Louis C. Karpinski.

813. Peter of Dacia. Petri Philomeni de Dacia in Algorismum vulgarem Johannis de Sacrobosco commentarius. Una cum Algorismo ipso edidit. Edited by Maximilian Curtze. Copenhagen: Høst, 1897. Pp. xix + 92.

814. Sacrobosco, John of. Tractatus de arte numerandi. Rara mathematica. 2d ed. Edited by J.O. Halliwell. London, 1841. Reprint. Hildesheim: Olms, 1977, pp. 1-26.

815. Schrader, Dorothy V. "De arithmetica, Book I of Boethius." Mathematics Teacher, 61 (1968), 615-628.

Surveys the content of Book I of this Boethian work, devoted to the properties and relationships of number, largely taken from Nicomachus of Gerasa.

816. Smith, David E. and Louis C. Karpinski. The Hindu-Arabic Numerals. Boston: Ginn and Company, 1911. Pp. vi + 160.

Traces the development of Hindu numerals in India, their transmission to and use by the Arabs and the avenues of introduction to Latin Europe.

817. Steele, Robert R., ed. The Earliest Arithmetics in English. Early English Text Society, extra ser., 118. London: Oxford University Press, 1922. Pp. xviii + 84.

Provides the texts of two English arithmetics, a translation and amplification respectively of the Carmen de algorismo of Alexander de Villedei and Sacrobosco's De arte numerandi, as well the texts of a work by Robert Record on the abacus (Accomptynge by Counters) and the Carmen de algorismo with material not in the version published by Halliwell. See item 771.

818. von Sterneck, R. Daublebsky. "Zur Vervollständigung der Ausgaben der Schrift des Jordanus Nemorarius: Tractatus de numeris datis." Monatschefte für Mathematik und Physik, 8 (1896), 165-179.

Contains additions to the text of the De numeris datis edited by Curtze. See item 781.

819. Vogel, Kurt, ed. Die Practica des Algorismus Ratisbonensis: Ein Rechenbuch des Benediktinerskloster St. Emmeram aus der Mitte des 15. Jahrhunderts. Munich: C.H. Beck, 1954. Pp. xi + 283.

An edition of the arithmetic problems in the Algorismus Ratisbonensis stemming from St. Emmeram at Regensburg, a center of mathematical interests in the late Middle Ages. The content of the Practica, written in Latin, German and a mixture of both, reveals mercantile and commercial interests characteristic of Italian influences. The main part of the work is due to Friedrich Gerhart.

820. Vogel, Kurt, ed. Ein italienisches Rechenbuch aus dem 14. Jahrhundert (Columbia X 511 A13). Munich: Deutsches Museum, 1977. Pp. vii + 186.

An edition of a text containing 142 arithmetical and geometrical problems by an author possibly from Cortona and intended for merchants. The mathematical content includes problems on the fundamental operations of arithmetic, those met with in the daily life of a merchant and mathematical riddles. The text shows the influence of Leonardo of Pisa and also introduces its readers to the Indian-Arabic numerals.

## GEOMETRY, PRACTICAL GEOMETRY, PROPORTIONS, CONIC SECTIONS, TRIGONOMETRY

821. Anaritius (al-Narīzī). Anaritii in decem libros priores Elementorum Euclidis commentarii. Ex interpretatione Gherardi Cremonensis. Edited by Maximilian Curtze. Leipzig: Teubner, 1899. Pp. xxix + 389.

822. Baron, Roger. "Sur l'introduction en occident des termes geometria theorica et practica." Revue d'histoire des sciences,, 8 (1955), 298-302.

Argues that Hugh of St. Victor, in his Practica geometriae, was the first to use these terms in a geometrical context. Geometry was a theoretical science. based on rational investigation, while practical geometry was devoted to measurement with instruments.The distinction, theorica-practica, was itself well-known. The former had as its object the contemplation of truth, while the latter treated that part of philosophy concerned with the moral and ethical life.

823. Busard, H.L.L. "The Practica geometriae of Dominicus de Clavasio." Archive for History of Exact Sciences, 2 (1965), 520-575.

Presents an edition of Dominicus' treatise on mensuration, written in 1346, consisting of an introduction and three chapters dealing respectively with the measurement of lengths, areas and volumes.

824. Busard, H.L.L. The Translation of the Elements of Euclid from the Arabic into Latin by Hermann of Carinthia (?). Leiden: Brill, 1968. Pp. 142.

Latin text of Books I-VI of the Elements. Concludes that the version in Bibliothèque Nationale lat. MS 16646 is probably due to Hermann of Carinthia. This translation was not used by Campanus. See item 830.

825. Busard, H.L.L. "Die Traktate De proportionibus von Jordanus Nemorarius und Campanus." Centaurus, 15 (1971), 193-227.

Gives editions of these two works of Jordanus and Campanus. Points out that the proportionality relationships of the theorem of Menelaus were used by two Arabic mathematicians, Thābit ibn Qurra and Ahmad ibn Jusuf (Ametus son of Joseph), and transmitted to the West. Campanus, in his treatise on proportions followed Ametus, while Jordanus used Thābit. However, neither Jordanus nor Campanus refer to the theorem and had other materials on proportions at their disposal. The ascription of a De proportionibus to Jordanus is questioned by Thomson. See item 1445.

826. Busard, H.L.L. "Über einige Euklid Kommentare und Scholien, die im Mittelalter bekannt waren." Janus, 60 (1973), 53-58.

Discusses briefly material on Books I and X of the Elements by Roger Bacon, on Books III and V in the De arcubus similibus and De proportione et proportionalitate of Ahmad ibn Jusuf, and in several scholia to Books X-XIV accompanying Gerard of Cremona's translation.

827. Busard, H.L.L. "Ein mittelalterlicher Euklid-Kommentar, der Roger Bacon zugeschreiben werden kann." Archives internationales d'histoire des sciences, 24 (1974), 199-217.

Believes there are reasons for ascribing this commentary in a Florence MS to Roger Bacon rather than Adelard of Bath. Provides the Latin text.

828. Busard, H.L.L. "The Second Part of Chapter 5 of the De arte mensurandi by Johannes de Muris." For Dirk Struik. Edited by R.S. Cohen, J.J. Stachel and M.W. Wartofsky. Dordrecht:Reidel, 1974, pp. 147-167.

Shows that the author of this section of the De arte mensurandi, dealing with the measurement of rectilinear figures, was influenced by,although did not draw directly from, the Liber mensurationum of the Arabic author Abū Bekr, whose work had been translated by Gerard of Cremona.

829. Busard, H.L.L. "Über einige Euklid-Scholien, die den Elementen von Euklid, übersetzt von Gerard von Cremona, angehängt worden sind." Centaurus, 18 (1974), 97-128.

Discusses and gives the text of these twenty-two *scholia*.

830. Busard, H.L.L. *The Translation of the Elements of Euclid from the Arabic into Latin by Hermann of Carinthia(?), Books VII-XII*. Amsterdam: Mathematisch Centrum, 1977. Pp. 198.

Latin text of the version in Biblithèque Nationale, lat. MS 16646, likely translated from Arabic by Hermann. Concludes that the Arabic version Hermann used was al-Ḥaggāg I, not that of Ishāq-Thābit. See item 824.

831. Busard, H.L.L. "Der Trakat *De isoperimetris*, der unmittelbar aus dem Grieschischen ins Lateinische übersetzt werden ist." *Mediaeval Studies*, 42 (1980), 61-88.

Argues that this anonymous translation of a work on isoperimetry, made directly from Greek, stems from an anonymous introduction to the *Almagest*, which may in turn, be an unknown part of the collection of treatises comprising the "little astronomy." The treatise shows a similarity to the *De curvis superficiebus* of Johannes de Tinemue, and was known to Roger Bacon, Bradwardine and Albert of Saxony.

832. Busard, H.L.L. and P.S. van Koningsveld. "Der *Liber de arcubus similibus* des Ahmed ibn Jusuf." *Annals of Science*, 30 (1973), 381-406.

Gives the Arabic and Latin text of this work by Ahmed (Ametus son of Joseph) which was likely translated by Gerard of Cremona.

833. Clagett, Marshall. "Archimedes in the Middle Ages: The *De mensura circuli*." *Osiris*, 10 (1952), 587-618.

Points out that the *De mensura circuli*, translated twice from Arabic (Plato of Tivoli and Gerard of Cremona), was frequently cited, leading to the conclusion that it was among the standard geometric works of the thirteenth and fourteenth centuries. See item 841.

834. Clagett, Marshall. "A Medieval Fragment of the *De sphaera et cylindro* of Archimedes." *Isis*, 43 (1952), 36-38.

Draws attention to a fragment of _De sphaera_ in an Oxford MS among other scientific works translated by Gerard of Cremona. See item 841.

835. Clagett, Marshall. "The Use of the Moerbeke Translations of Archimedes in the Works of Joannes de Muris." _Isis_, 43 (1952), 236-242.

Establishes that John of Murs used a hybrid collection of Archimedean works consisting of parts of _The Measurement of a Circle_ and _On Spiral Lines_ called _De quadratura circuli_, dated about 1340, in his _Quadripartitum_, and joined it to the eighth chapter of his _De mensurandi ratione_. Gives a detailed analysis of _De quadratura._ See item 848, pt. I.

836. Clagett, Marshall. "The _De curvis superficiebus Archimenidis_: A Medieval Commentary of Johannes de Tinemue on Book I of the _De sphaera et cylindro_ of Archimedes." _Osiris_, 11 (1954), 295-358.

Establishes that this popular work in the Archimedean tradition, inspired by Book I of _De sphaero et cylindro_ and _De mensura circuli_, was composed in the early thirteenth century. The purpose of the work is to determine the surface areas and volumes of certain solids bounded by curved surfaces. The text is included and an addendum provides further propositions. See item 841.

837. Clagett, Marshall. "A Medieval Latin Translation of a Short Arabic Text on the Hyperbola." _Osiris_, 11 (1954), 359-385.

Provides the text with translation of a work translated into Latin by John of Palermo who was at the court of Frederick II. Relying mainly on Apollonius, its purpose was to demonstrate the asymtotic property of the hyperbola. See item 849, pt. I.

838. Clagett, Marshall. "The _Quadratura per lunulas_: A Thirteen-Century Fragment of Simplicius' Commentary on the _Physics_ of Aristotle." _Essays on Medieval Life and Thought in Honor of Austin Patterson Evans._ Edited by John H. Mundy, Richard W. Emery and Benjamin N. Nelson. New York: Columbia University Press, 1955, pp. 99-108.

Discusses and gives the text of two versions of a proposition containing an incorrect quadrature of the circle by the quadrature of lunes reported by Simplicius from Alexander of Aphrodisias. One version is a translation, possibly by Grosseteste; the other, more popular, is a paraphrase which omits Simplicius' comment that the argument is fallacious. See item 848, pt. IV.

839. Clagett, Marshall. "A Medieval Treatment of Hero's Theorem on the Area of a Triangle in Terms of its Sides." Didascaliae: Studies in Honor of Anselm M. Albareda. Edited by Sesto Prete. New York: Rosenthal, 1961, pp. 79-95.

Points out that Hero's theorem for the area of a triangle entered the Latin West with Gerard of Cremona's translation of the Verba filiorum of the Banū Mūsā. A different proof of the theorem, possibly of Arabic, Greek or Latin origin, and associated with Jordanus de Nemore also circulated; two versions of the latter are edited here. See item 841.

840. Clagett, Marshall. "Archimedes and Scholastic Geometry." Mélanges Alexandre Koyré, I. Paris: Hermann, 1964, pp. 40-60.

Presents the various changes that took place in the presentation and proof of the problem of the quadrature of the circle during the late thirteenth and fourteenth centuries.

841. Clagett, Marshall. Archimedes in the Middle Ages. Vol. I: The Arabo-Latin Tradition. Madison: University of Wisconsin Press, 1964. Pp. xxix + 720.

Traces by an analysis of the available Latin texts (with translations), the advent and use made of Archimedean treatises, particularly De mensura circuli and De sphaera et cylindro, and of materials based on Archimedean methods and results, which entered the West through translation from Arabic before Moerbeke's Greek versions of the Archimedean corpus. An exception included is the Liber de curvis superficiebus of Johannes de Tinemue which was translated from Greek.

842. Clagett, Marshall. "A Medieval Archimedean-Type Proof of the Law of the Lever." Studies in Medieval Physics

and Mathematics. London: Variorum Reprints, 1979, pp. 409-421.

Provides the text of a proof from a MS at the Bibliothèque Nationale, Paris, probably of the fourteenth century, which substitutes a proof of an Archimedean kind for the dynamic version of Jordanus. See item 848, pt. I. Reprinted from Miscellanea André Combes, II. Rome, 1967.

843. Clagett, Marshall. "Johannes de Muris and the Problem of Proportional Means." Medicine, Science and Culture: Historical Essays in Honor of Owsei Temkin. Edited by Lloyd G. Stevenson and Robert P. Multhauf. Baltimore, Johns Hopkins University Press, 1968, pp. 35-49.

Maintains that Johannes' treatment of finding two proportional means between two given quantities, found in his De arte mensurandi (Chapter 7, Proposition 16), stems from Moerbeke's translation of Eutocius' commentary on Archimedes' On the Sphere and the Cylinder. See item 848, pt. I.

844. Clagett, Marshall. "Leonardo da Vinci and the Medieval Archimedes." Physis, 11 (1969), 100-151.

Examines the notebooks of Leonardo for direct and indirect traces of works of Archimedes. In several cases, Leonardo's knowledge originated in medieval sources or in Renaissance works based on medieval sources. Leonardo saw copies of both the medieval translation by Moerbeke (1269) and the Renaissance translation of Jacobus Cremonensis (ca. 1450), although his direct use of these versions was limited primarily to On the Equilibrium of Planes. See item 848, pt. III.

845. Clagett, Marshall. "A Note on the Commensurator Falsely Attributed to Regiomontanus." Isis, 60 (1969), 383-384.

Establishes that this work is a collection of propositions from the fourteenth-century De arte mensurandi of John de Muris.

846. Clagett, Marshall. "The Quadrature by Lunes in the Later Middle Ages." Philosophy, Science, and Method:

*Essays in Honor of Ernst Nagel*. Edited by Sidney Morgenbesser, Patrick Suppes and Morton White. New York: St. Martin's Press, 1969, pp. 508-522.

Shows in addition to the conventional solution of Archimedes to the quadrature of the circle, two versions of an erroneous solution based on the quadrature of lunes also circulated in the late Middle Ages. See item 848, pt. IV.

847. Clagett, Marshall. *Archimedes in the Middle Ages*. Vol. II. *The Translations from the Greek by William of Moerbeke*. 3 parts in 2 vols. *Memoirs of the American Philosophical Society*, 117A-B. Philadelphia: American Philosophical Society, 1976. Pp. 698.

Texts of Archimedean works translated by Moerbeke.

848. Clagett, Marshall. *Archimedes in the Middle Ages*. Vol. III: *The Fate of the Medieval Archimedes, 1300-1565*. 4 parts in 3 vols. *Memoirs of the American Philosophical Society*, 125A-C. Philadelphia: American Philosophical Society, 1978. Pp. 1582.

Part I discusses the use of the Moerbeke translations during the fourteenth century. Part II treats the fortunes of the Arabic Archimedean translations in this century and surveys the handbooks on practical geometry influenced by Roman and Arabic gromatic materials and by the Archimedean tradition. Part III traces the continued use of the Moerbeke translations and other versions of Archimedes by Renaissance authors. Part IV contains appendixes and diagrams.

849. Clagett, Marshall. *Archimedes in the Middle Ages*. Vol. IV: *Conic Sections*. 2 parts in 2 vols. *Memoirs of the American Philosophical Society*, 137A-B. Philadelphia: American Philosophical Society, 1980. Pp. 566.

Supplementary volumes on the medieval Latin tradition of conic sections (1150-1566). Argues that the Archimedean and Eutocian treatment in Moerbeke's translations exercised little influence, but that conics did receive attention as part of the optical tradition. In the late Middle Ages there was an increasing interest in optics and this medieval tradition continued through the first half of the sixteenth century.

850. Clagett, Marshall. "Conic Sections in the Fourteenth Century: The Nature of the Latus Rectum of a Parabola." Studi sul XIV secolo in memoria Anneliese Maier. Edited by Alfonso Maierù and Agostino Paravicini Bagliani. Rome: Edizioni di storia e letteratura, 1981, pp. 179-217.

Analyzes the treatment of the latus rectum of a parabola in the Speculi almukefi compositio attributed to Roger Bacon (early fourteenth century), and the Libellus de seccione of Jean Fusoris (late fourteenth century). See item 849, pt. I.

851. Curtze, Maximilian. "Practica geometriae." Monatshefte für Mathematik und Physik, 8 (1897), 193-224.

Provides the Latin text with commentary of a Practica of the twelfth century.

852. Curtze, Maximilian. "Eine Studienreise." Centralblatt für Bibliothekswesen, 16 (1899), 264-265.

Gives the text of the propositions of an Isoperimetria ascribed to Jordanus. This ascription is considered dubious by Thomson. See item 1445.

853. Curtze, Maximilian. "Urkunde zur Geschichte der Trigonometrie im christlicher Mittelalter." Biblioteca mathematica, ser. 3, 1 (1900), 321-416.

Provides selections from MSS of trigonometry in texts by Savasorda, Arzachel, Levi ben Gerson, Johannes de Lineriis and Johannes de Muris.

854. Drecker, J. "Das Planisphaerium des Claudius Ptolemaeus." Isis, 9 (1927), 255-278.

The Latin text of Ptolemy's work on stereographic projection as translated in 1143 by Hermann of Carinthia (Dalmatia) from Arabic.

855. Espenshade, Pamela H. "A Text on Trigonometry by Levi ben Gerson (1288-1344)." Mathematics Teacher, 60 (1967), 628-637.

Translation with notes and commentary of De sinibus, chordis et arcubus.

856. Euclid. The Medieval Latin Translation of the Data of Euclid. Edited by Shuntaro Ito. Boston: Birkhäuser, 1980. Pp. ix + 256.

857. Evans, Gillian R. "The 'sub-Euclidean' Geometry of the Earlier Middle Ages, up to the Mid-Twelfth Century." Archive for History of Exact Sciences, 16 (1976), 105-118.

Treats the fortunes of a simple geometry not directly derived from Euclid, found in early medieval commentaries, handbooks, encyclopedias, Augustine, and treatises on practical geometry--a "sub-Euclidian" geometry, which was widely read by many scholars. Although the Elements was available by the twelfth century, this "sub-Euclidean" tradition persisted.

858. Evans, Gillian R. "Boethius' Geometry and the Four Ways." Centaurus, 25 (1981), 161-165.

Suggests the topics Boethius might have discussed in his lost Geometria based on his two surviving quadrivial treatises, the Musica and the Arithmetica. His Geometria would have included a translation of Euclid, but also criticism of the text, and would have contained philosophical speculations and references to the other quadrivial sciences.

859. Folkerts, Menso. "Das Problem der pseudo-Boethischen Geometrie." Sudhoffs Archiv: Vierteljahrsschrift für Geschichte der Medizin, der Naturwissenschaften, der Pharmazie und der Mathematik, 52 (1968), 152-161.

Discusses the nature and content of one of the two versions of the Boethian translation of Euclid, the so-called "Boethius" Geometry II, consisting of two books on geometry and arithmetic divided into three sections: on the abacus, extracts from the agrimensores and excerpts from Euclid. The part on the abacus was taken from a version of Gerbert's Regulae de numerorum abaci rationibus. The Euclid excerpts are similar to those in "Boethius" Geometry I. The mathematical competence of the unknown author was meager; he composed the work after 1000 and probably lived in Lorraine.

860. Folkerts, Menzo. "Anonyme lateinische Euklidbearbeitungen aus dem 12. Jahrhundert." _Österreichische Akademie der Wissenschaften, math.-natur. Klasse_. _Denkschriften_, 116, 1 (1971), 5-42.

Maintains that while the Latin Euclid tradition can be divided into materials derived from translations from either Greek or Arabic, not all texts of the medieval Euclid can be placed with certainty in one group or the other. Some versions, which are discussed here, are mixed, revealing the interaction of the two segments of the tradition. As an example, the text of an anonymous _Geometria_ attributed to Boethius is compared with the text of Adelard, version II.

861. Folkerts, Menso. "Probleme der Euklidinterpretation und ihre Bedeutung für die Entwicklung der Mathematik." _Centaurus_, 23 (1980), 185-215.

Traces the fortunes of the _Elements_ in antiquity, its transmission to and fate among the Arabs, the routes of its transfer to the Latin West and its subsequent treatment in the sixteenth century. Provides an example in the treatment of the problem of proportionality arising from Book V from late antiquity through the sixteenth century.

862. Folkerts, Menzo. "The Importance of the Pseudo-Boethian _Geometria_ during the Middle Ages." _Boethius and the Liberal Arts_. Edited by Michael Masi. Berne: Lang, 1981, pp. 187-209.

Traces the fortunes of the Boethian translation of Euclid, which while not available in the original, has partially survived in two compilations, both enlarged by the addition of gromatic and arithmetic materials, the _Geometria_ I, connected with eighth-century Corbie and the _Geometria_ II, which originated in eleventh-century Lorraine. Both of these compilations were popular sources of geometrical information until the return of the complete Euclid and other Greek mathematical materials in the twelfth century. The attempt made at this time to combine the earlier Greek Euclid with the newer Arabic-Latin tradition proved short-lived. Neither the _Geometria_ I or II was of importance in the late medieval period, although interest in them revived with Renaissance humanists.

863. Franco of Liège. "Der Traktat Franco's von Luettich De quadratura circuli." Edited by Dr. Winterberg. Abhandlungen zur Geschichte der Mathematik, 4 (1882), 137-190.

864. Franco of Liège. "A Treatise on the Squaring of the Circle by Franco of Liège, of about 1050." Edited by Menso Folkerts and A.J.E.M. Smeur. Archives internationales d'histoire des sciences, 26 (1976), 59-105; 225-253.

Part I of this article presents a new edition of the treatise while Part II contains a detailed analysis. Franco, who became head of the famous school at Liège in 1066, probably composed this work around 1050. The editors point out that, besides his attempt to square the circle, Franco's treatise deals primarily with irrational square roots. They maintain that he was not only familiar with the mathematical knowledge of his time, but approached these problems in a systematic and commendable way, considering that he had no direct access to Greek mathematics.

865. Freguglia, Paola. "Osservazioni sul procedimento di Leonardo Pisano per la determinazione di pi." Cultura e scuola, 13 (1974), 209-221.

Compares Leonardo's procedure for calculating pi (in a passage in his Practica geometriae) with that of Archimedes.

* Hahn, Nan L. Medieval Mensuration: Quadrans vetus and Geometrie due sunt partes principales.... Transactions of the American Philosophical Society, 72, Part 8. Philadelphia: American Philosophical Society, 1982. Pp. lxxxv + 204.

Cited as item 544.

866. Hugh of St. Victor. "Hugonis de Sancto Victore Practica geometria." Edited by Roger Baron. Osiris, 12 (1956), 176-224.

Establishes Hugh as the author of this treatise and provides an edition of the text.

867. Hughes, Barnabas, trans. Regiomontanus on Triangles. Madison: University of Wisconsin Press, 1967. Pp. vii + 298.

Contains the text and English translation of the De triangulis omnimodis, a work on trigonometry. The text is taken from the first edition published at Nuremberg for John Schöner in 1533.

868. John of Gmunden. "Der Traktat De sinibus, chordis et arcubus von Johannes von Gmunden." Edited by H.L.L. Busard. Österreichische Akademie der Wissenschaften, math.-natur. Klasse. Denkschriften, 116, 3 (1971), 73-113.

An edition of John's treatise. In this work, written in 1437, John of Gmunden is revealed as the first of the fifteenth-century scholars at Vienna to introduce new trigonometric treatises.

869. Jordanus de Nemore. Jordani Nemorarii Geometria vel De triangulis libri IV. Edited by Maximilian Curtze. Mitteilungen des Coppernicus-Vereins für Wissenschaft und Kunst zu Thorn, 6. Thorn: Lambeck, 1887. Pp. xv + 50.

870. Jordanus de Nemore. Jordanus de Nemore and the Mathematics of Astrolabes: De plana spera. Edited with introduction, translation and commentary by Ron B. Thomson. Toronto: Pontifical Institute of Mediaeval Studies, 1978. Pp. x + 238.

Provides three versions of the treatise on stereographic projection, the mathematical basis of astrolabe construction.

871. Lull, Ramón. "De quadratura et triangulatura circuli." Edited by J.E. Hofmann. Cusanus-Studien, VII. Die Quellen der Cusanischen Mathematik, I: Ramon Lulls Kreisquadratur. Sitzungsberichte der Heidelberger Akademie der Wissenschaften, phil.-hist. Klasse, 4 (1942), pp. 1-37.

Maintains that Lull's treatise was a direct source for Nicholas of Cusa; there is a copy in the library the latter established at Cues. Lull's work is not that of a mathematician; he used mathematics, as did Cusa, to find a deeper, speculative and mystical meaning of general philosophical significance.

872. Lull, Ramón. El libro de nova geometria. Edited with introduction and notes by J.M. Millás Vallicrosa. Barcelona: Asociación para la historia española, 1953. Pp. 104.

See item 212.

873. Mainardi, Leonardo. "Die Practica geometriae des Leonardo Mainardi aus Cremona." Edited by Maximilian Curtze. Urkunden zur Geschichte der Mathematik im Mittelalter und der Renaissance. 2 vols. in 1. Leipzig, 1902. Reprint. New York: Johnson Reprint, 1968, pp. 337-434.

Edition of the Artis metrice practice compilatio, extant in Italian; text with German translation.

874. Mendthal, H., ed. Geometria culmensis: Ein agronomischer Tractat. Verein für Ost-und Westpreussen. Leipzig: Duncker & Humblot, 1886. Pp. 76.

An anonymous work containing problems in field measurement written at the demand of Conrad von Jungingen, grand-master of the Teutonic Knights, 1393-1407. The work is extant in both Latin and medieval German; both versions are provided.

875. Molland, A.G. "An Examination of Bradwardine's Geometry." Archive for History of Exact Sciences, 19 (1978), 113-175.

Provides a synopsis and analysis of Bradwardine's Geometria speculativa. Points out such characteristics of the work as its dependency on the Campanus version of Euclid, its philosophical emphasis which reveals Bradwardine as a realist. Moreover, written in a scholastic style, it does not treat constructions in a satisfactory way nor is it concerned with geometric rigor.

876. Murdoch, John E. "The Medieval Language of Proportions: Elements of the Interaction with Greek Foundations and the Development of New Mathematical Techniques." Scientific Change. Edited by A.C. Crombie. New York: Basic Books, 1963, pp. 237-271.

Discusses the Eudoxian-Euclidean foundations of proportionality theory, tracing the fortunes of Book V, Def. 4 (continuity principle) and Book V, Def. 5 (equality of proportions) of the Elements among Arabic mathematicians and in the Latin West. In the West, the status of V, Def. 4 was affected by a lack of the genuine definition; in the case of V, Def. 5, the definition was transmitted correctly, but was misunderstood. Despite difficulties with the Greek heritage, medieval authors displayed great interest in the theory of proportions and proportionality and extended its application to mechanics.

877. Murdoch, John E. "Superposition, Congruence, and Continuity in the Middle Ages." Mélanges Alexandre Koyré, I. Paris: Hermann, 1964, pp. 416-441.

Traces the use made of one mathematical notion, that of superposition, the congruence of geometrical figures, in scholastic interpretations of the continuum as an example of the role played by mathematical ideas and methods by schoolmen.

878. Oresme, Nicole. Quaestiones super geometriam Euclidis. Edited by H.L.L. Busard. Leiden: Brill, 1961. Pp. xiv + 179.

Points out that the scholastic format of the work suggests it was composed when Oresme was teaching at Paris. Questions 10-17 present the graphical representation of qualities and motions more fully treated in the De configurationibus. Points out the resemblance between questions 6-9 and the De proportione dyametri quadrati ad costam ejusdam Suter attributed to Albert of Saxony. See items 884, 895.

879. Oresme, Nicole. "Part 1 of Nicole Oresme's Algorismus proportionum." Edited and annotated by Edward Grant. Isis, 56 (1965), 327-341.

Presents an English translation of Oresme's Algorism of Ratios, the first known systematic attempt to provide operational rules for the multiplication and division of ratios involving integral and fractional exponents.

880. Oresme, Nicole. De proportionibus proportionum and Ad pauca respicientes. Edited with introductions,

English translations and critical notes by Edward Grant. Madison: University of Wisconsin Press, 1966. Pp. xxii + 466.

Points out that the primary contribution of the *De proportionibus* is Oresme's distinction between irrational ratios with rational exponents and those with irrational exponents. Asserts that the *Ad pauca respicientes* (named for its opening words) is a preliminary version of the *De commensurabilitate*. See item 480.

881. Paravicini Bagliani, Agostino. "Un matematico nella corte papale del secolo XIII: Campano de Novara (d. 1296)." *Rivista di storia della chiesa in Italia*, 27 (1973), 98-129.

Presents information summarizing what is known of Campanus' origins and family connections. Argues that the reconstruction of his biography is bound up with the history of the papal court during the late thirteenth century since most of the available documents are linked to his 30-year stay at the curia.

882. Savasorda (Abraham bar Ḥiyya). "Der *Liber embadorum* des Savasorda in der Übersetzung des Plato von Tivoli." Edited by Maximilian Curtze. *Urkunden zur Geschichte der Mathematik im Mittelalter und der Renaissance*. 2 vols. in 1. Leipzig, 1902. Reprint. New York: Johnson Reprint, 1968, pp. x + 183.

Text with German translation.

883. Suter, Heinrich. "Der Tractatus *De quadratura circuli* des Albertus de Saxonia." *Zeitschrift für Mathematik und Physik*, *hist.-lit. Abteilung*, 29 (1884), 81-101.

Latin text of the treatise.

884. Suter, Heinrich. "Die *Questio de proportione dyametri quadrati ad costam ejusdem* des Albertus de Saxonia." *Zeitschrift für Mathematik und Physik*, *hist.-lit. Abteilung*, 32 (1887), 41-56.

Latin text of the work. The attribution to Albert of Saxony has been challenged. See items 878, 894, 895.

885. Tannery, Paul. "Une correspondance d'écolâtres du onzième siècle." Mémoires scientifiques, V. Issued by J.-L. Heiberg. Paris: Gauthier-Villars, 1922, pp. 103-111, 229-303.

Provides the text of and comments on the exchange of letters between Ragimbold of Cologne and Radolf of Liège (ca. 1025). The subject of the letters is geometry, although one of them (from Radolf) mentions an astrolabe which may have been furnished by Fulbert of Chartres. This correspondence reveals the geometrical knowledge of and the level of geometrical teaching in northern Europe before 1050. See item 1299.

886. Trummers, Paul M.J.E. "The Commentary of Albert on Euclid's Elements of Geometry." Albertus Magnus and the Sciences: Commemorative Essays. Edited by James A. Weisheipl. Toronto: Pontifical Institute of Mediaeval Studies, 1980, pp. 479-499.

Argues that internal evidence and correspondences with authentic works lead to the conclusion that Albert did write a commentary on Book I of Euclid, likely between 1235-1260. Surveys the commentary, pointing out its reliance on Anaritius and several versions of Euclid; concludes that the work does not reveal a high level of mathematical insight.

887. Ullman, B.L. "Geometry in the Mediaeval Quadrivium." Studi di bibliografia e di storia in onore di Tammaro de Marinis, IV. Edited by Giovanni Mardersteig. Verona: Valdonega, 1964, pp. 263-285.

Focuses on the fortunes of the agrimensores-literature from the ninth to the eleventh centuries, arguing that the history of this tradition can be traced through extant MSS and library catalogues, that treatises on gromatic arts were a source of material for the theoretical geometry taught in the quadrivium and that the foundation at Corbie was a center for both geometry and the gromatic arts.

888. Unguru, Sabetai. "Witelo and Thirteen-Century Mathematics: An Assessment of his Contribution." Isis, 63 (1973), 496-508.

Discusses the first book of Witelo's Perspectiva, a compilation of different geometrical theorems Witelo thought necessary to understand the remaining nine optical books, and the sources Witelo utilized. While Witelo was not an outstanding mathematician, in the context of the mathematical aridity of the thirteenth century, he showed an uncommon knowledge and understanding of geometry.

889. Unguru, Sabetai. "Pappus in the Thirteenth Century in the Latin West." Archive for History of Exact Sciences, 13 (1974), 307-324.

Refutes the view that Pappus of Alexandria's Collectio mathematica was unknown in the Latin Middle Ages as Witelo in his Perspectiva (1270's) knew at least part of it.

890. Unguru, Sabetai. "A Very Early Acquaintance with Apollonius of Perga's Treatise on Conic Sections in the Latin West." Centaurus, 20 (1976), 112-128.

Argues that the detailed knowledge of Apollonius' Conics in Witelo's Perspectiva must have come from a Latin translation subsequent to the fragmentary twelfth-century version from Arabic of Gerard of Cremona. A probable conjecture is that such a translation was made by Witelo's friend, William of Moerbeke.

891. Unguru, Sabetai. "A Thirteenth-Century 'Proof' of the Parallel Postulate." Historia mathematica, 5 (1978), 205-210.

Points out that the "proof" of Euclid's fifth postulate in Book I of Witelo's Perspectiva is a petitio principii. Book I contains geometrical material supposed to serve as a foundation for the subsequent optical books. Argues that Witelo,as have many other mathematicians, may have been dissatisfied with the parallel postulate's status as a postulate not requiring proof.

892. Victor, Stephen. "Johannes de Muris' Autograph of the De arte mensurandi." Isis, 61 (1970), 389-395.

Argues that the fourteenth-century copy of this treatise in Bibliothèque National MS lat. 7380 is a de Muris autograph.

893. Victor, Stephen, ed. and trans. Practical Geometry in the High Middle Ages: Artis cuiuslibet consummatio and the Pratike de geometrie. Memoirs of the American Philosophical Society, 134. Philadelphia: American Philosophical Society, 1979. Pp. xii + 638.

An edition of a late twelfth-century practical geometry and its thirteenth-century French adaptation. Introduces the edition by a survey of practical geometry in the works of classifiers of the sciences, its relation to theoretical geometry, its role in medieval education and its function as a basis for surveying, city planning and architecture.

894. Zoubov, V.P. "Quelques observations sur l'auteur du traité anonyme Utrum dyameter alicuius quadrati sit commensurabilis costae ejusdem." Isis, 50 (1959), 130-134.

Argues that the treatment of the kinematics of circular motion in this treatise which Suter had attributed to Albert of Saxony reveals many similarities to both Oresme's De commensurabilitate and his De proportionibus and may be by him. See item 884.

895. Zoubov, V.P. "Autour des Quaestiones super geometriam Euclidis de Nicole Oresme." Mediaeval and Renaissance Studies, 6 (1968), 150-172.

Points out the similarities in passages dealing with the question of the commensurability of the diagonal of a square to its side in Busard's edition of the Quaestiones and the treatise published by Suter which Zoubov has ascribed to Oresme. See items 884, 894.

# MEDICINE AND VETERINARY MEDICINE

## THEORY AND PRACTICE

896. Alderotti, Taddeo. Thaddaeus Florentinus, I consilia. Edited by Giuseppe M. Nardi. Turin: Minerva medica, 1937. Pp. xlviii + 242.

  Presents the Consilia of Alderotti in the context of medical teaching at Bologna.

897. Arnald of Villanova. Opera medica omnia. 19 vols. Granada: Seminarium historiae medicae Granatensis, 1975.

898. Arnald of Villanova. Aphorismi de gradibus. Edited by Michael R. McVaugh. Opera medica omnia,, II. Granada: Seminarium historiae medicae Granatensis, 1975. Pp. xiv + 338.

  See item 897.

899. Averroes (Ibn Rushd). Colliget. Venice, 1962-1574. Reprint. Frankfort: Minerva, 1962. Pp. 312.

  See item 3.

900. Avicenna (Ibn Sina). Liber canonis. Venice, 1507. Reprint. Hildesheim: Olms, 1964.

  Translated by Gerard of Cremona.

901. Bacon, Roger. Fratris Rogeri Bacon De retardatione accidentium senectutis cum aliis opusculis de rebus

medicinalibus. Edited by A.G. Little and E. Withington. Oxford, 1928. Reprint. Farnborough: Gregg, 1966. Pp. xliv + 224.

902. von Baierland, Ortolf. Das Arzneibuch. Edited by James Follan. Stuttgart: Wissenschaftliche Verlagsgesellschaft, 1963. Pp. 199.

903. Bercovy, David. "Un nom illustre parmi les médecins juifs de Provence: Ibn-Tibbon." Revue d'histoire de la médecine hébräique, 20 (1967), 169-176.

Treats the Tibbonides family who flourished in southern France at the end of the eleventh and beginning of the twelfth century, whose best known representative was the astronomer known in the Latin West as Profatius Judaeus.

904. Bernard of Gordon. Il De urinis di Bernardo di Gordon (XIII-XIV sec.). Edited and translated by Pier Luigi Mondani. Scientia veterum, 128. Pisa: Giardini, 1968. Pp. 128.

Gives only the Italian translation of the text.

905. Bietti, Gianbattista. "Le influenze dell'oftalmologia araba sulla oftalmologia medioevale Europea." Oriente e occidente nel medioevo: Filosofia e scienze. Rome: Accademia nazionale dei Lincei, 1971, pp. 445-452.

Argues that Arabic medical writers transcended their Greek sources, basing their writings on direct experience. They devoted much attention to diseases of the eyes, motivated by the prevalence of such maladies in the Arabic world. This development is exemplified by the work of Hunain ibn Isḥāq on trachoma. Beginning with the translations of Constantine at Salerno, medieval Europe found in the Arabs their primary authorities on eye diseases.

906. Bonser, Wilfred. "General Medical Practice in Anglo-Saxon England." Science, Medicine and History: Essays on the Evolution of Scientific Thought and Medical Practice written in Honour of Charles Singer, I. Edited by E. Ashworth Underwood. London: Oxford University Press, 1953, pp. 154-163.

Characterizes the major influences in Anglo-Saxon medicine from the mid-fifth to the mid-eleventh century as coming from the church, classical sources, and Germanic folk traditions. Cures were often sought through the intervention of saintly relics or by the exorcism of demons. Folk tradition provided an element of fantasy and magic; however, the large number of recipies found in Anglo-Saxon herbals and leechbooks reveals a rational component.

907. Bonser, Wilfred. The Medical Background of Anglo-Saxon England: A Study in History, Psychology and Folklore. London: Wellcome Historical Medical Library, 1963. Pp. xxxv + 448.

Surveys medical thought and practice, discussing the extent and character of medical knowledge and the treatment of the sick, sources of medical information, reaction to epidemics, surgery, diet, veterinary and agricultural magical practices.

908. Bruce-Chwatt, L.J. "A Medieval Glorification of Disease and Death." Medical History, 16 (1972), 76-77.

Contains a translation of a work by the Italian Jacopo de Benedetti.

909. Bullough, Vern L. "Medieval Medical and Scientific Views of Women." Viator, (1973), 483-501.

Argues that clerical misogyny in the Middle Ages was re-enforced by physiological and anatomical notions regarding women inherited from antiquity.

910. Bullough, Vern L. and Cameron Campbell. "Female Longevity and Diet in the Middle Ages." Speculum, 55 (1980), 317-325.

Argues that the thirteenth-century rise in the longevity of women may be due to improvements in the medieval diet, particularly in the consumption of iron-rich foods which (in the light of women's greater need for iron), reduced the prevalence of iron-deficiency anemia as a contributing cause of death in women.

911. Cholmeley, Henry P. John of Gaddesden and the Rosa medicinae. Oxford: Clarendon Press, 1912. Pp. 184.

Provides information as to Gaddesden's probable studies in arts and medicine at Oxford, and discusses some of the topics in his Rosa Anglica of 1314. There is a translation of the Isagoge of Joannitius in an appendix.

912. Clay, Mary Rotha. The Mediaeval Hospitals of England. London, 1909. Reprint. New York: Barnes and Noble, 1966. Pp. xxii + 357.

Discusses the hospital as an ecclesiastical institution including foundations for pilgrims, the elderly, the needy and lepers, as well as the physically and mentally ill. Treats many aspects of the hospital--founders, organization, officials, care provided and categories of inmates. The period covered ends at 1547; there is a list of hospitals in an appendix. Those which were an integral part of a monastic house and those of the orders of the Temple and of St. John are excluded.

913. Codellas, P.S. "The Case of Smallpox of Theodorus Prodromus (XII Century A. D.)." Bulletin of the History of Medicine, 20 (1946), 207-215.

Points out that Prodromos, himself afflicted by smallpox, described the disease in detail in his correspondence, providing the earliest case of smallpox described by a European.

914. Demaitre, Luke E. "Theory and Practice in Medical Education at the University of Montpellier in the Thirteenth and Fourteenth Centuries." Journal of the History of Medicine and Allied Sciences, 30 (1975), 103-123.

Argues based on materials from Montpellier that there was an interplay between theory and practice in this medical faculty. Scholastic discussions had implications for medical practice while the change toward a concentration on the theoretical at Montpellier in the fourteenth century is not definite. There were fourteenth-century treatises on practical medicine.

915. Demaitre, Luke E. "Nature and the Art of Medicine in the Later Middle Ages." Mediaevalia, 2 (1976), 23-47.

Discusses attitudes toward nature in relation to medicine, a legacy from antiquity, as seen in medical sources from late thirteenth- and fourteenth-century France, focusing on implications for medical theory, treatment, causes of disease, the prolongation of life and ethics.

916. Demaitre, Luke E. "Scholasticism in Compendia of Practical Medicine, 1250-1450." Science, Medicine and the University: 1200-1550. Essays in Honor of Pearl Kibre, I. Edited by Nancy G. Siraisi and Luke Demaitre. Manuscripta, 20 (1976), pp. 81-95.

Argues that one must define the term "scholastic" with care in classifying the practical compendia or summas which summarize the practicing physician's procedures and which were not intended for university instruction. These works are quite diverse and are not uncritical of traditional authority; some show independence of thought and the results of personal experience.

917. Demaitre, Luke E. Doctor Bernard de Gordon: Professor and Practitioner. Toronto: Pontifical Institute of Mediaeval Studies, 1980. Pp. xii + 236.

Analyzes the thought of Bernard, whose career was spent in the teaching and practice of medicine at Montpellier, by an extensive treatment of his writings which cover the full medical curriculum. Concludes that despite his reliance on medical authorities and his limited clinical observations, Bernard used his sources in a critical fashion.

918. Diepgen, Paul. "Studien zu Arnald von Villanova." Archiv für Geschichte der Medizin, 3 (1909-10), 115-130, 188-196, 369-396; 5 (1911-12), 88-120; 6 (1912-13), 380-391. Reprint. Weisbaden: Steiner, 1964.

Discusses Arnald's education and career, the question of the genuineness of both the Brevarium and the De sterilitate, Arnald's supposed role as an alchemist and his position on magic, astrology and oneiromancy.

919. Diepgen, Paul. Frau und Frauenheilkunde in der Kultur des Mittelalters. Stuttgart: Thieme, 1963. Pp. viii + 242.

Traces the fortunes of medical theories and anatomical information concerning women's diseases through the medieval period, including Islam, relating gynecological practice to attitudes toward women in varying social circumstances. The treatment of women's ailments was heavily indebted to information stemming from antiquity, although this was modified by observation and experience. Many conditions peculiar to women were in the hands of midwives and "wise" women, and only slowly passed under control of professional male doctors.

920. Dureau, Jeanne-Marie. "Mathématique et médecine au XVe siècle: L'oeuvre du médecin catalan Antoine Ricart." Bulletin philologique et historique du comité des travaux historiques et scientifiques, (1964), 833-843.

Discusses the medical work of this virtually unknown Catalan physician (d. 1422). Moved by a strong desire to make medicine an exact science, Ricart extended the system of quantification of medical degrees based on Galen, Alkindi and Arnald of Villanova not only to medicines but to climates and humors. In his treatise on blood-letting, he adapted this system in an original fashion by treating the equilibrium or disequilibrium of humors in a statistical way. See item 921.

921. Dureau-Lapeyssonnie, Jeanne-Marie. "L'oeuvre d'Antoine Ricart, médecin catalan du XVe siècle: Contribution à l'étude des tentatives médiévales pour appliquer les mathématiques à la médecine." Médecine humaine et vétérinaire à la fin du moyen âge. Edited by Guy Beaujouan. Paris: Minard, 1966, pp. 169-364.

Treats Ricard's life, career and work, and analyzes the content of his Compendium secundi operis de arte graduandi medicinae compositas and his De quantitatibus et proportionibus humorum or De bona habitudine. (This latter work is edited in Appendix II). Ricard applied a theory of mathematical degrees to both doses of medicines and to combinations of humors.

* Eastwood, Bruce S. The Elements of Vision: The Micro-Cosmology of Galenic Visual Theory According to

Hunayn ibn Isḥāq. Transactions of the American Philosophical Society, 72, part 5. Philadelphia: American Philosophical Society, 1982. Pp. 58.

Cited as item 390.

922. Gratton, J.H.G. and Charles Singer. Anglo-Saxon Magic and Medicine. London: Oxford University Press, 1952. Pp. xii + 234.

A study divided into two parts: A survey of medical practice as employed by Anglo-Saxon leeches, stressing magical elements, and an analysis of the sources of Anglo-Saxon medical/magical writings. Part two is an edition with translation of Lacnunga, an Anglo-Saxon "commonplace" book containing recipes for the treatment of various ailments mixed with charms, incantations and prayers of pagan and Christian origin.

923. Gruner, O. Cameron. A Treatise on the Canon of Medicine of Avicenna. London, 1930. Reprint. New York: AMS Press, 1973. Pp. vii + 612.

An English translation of Book I.

924. Guerrino, Antonio A. and Alfredo G. Kohn Longarica. "La uroscopia en la edad media." Episteme, 7 (1973), 289-297.

Focuses on the importance of urine as a diagnostic method, in use from antiquity through the Renaissance, as revealed not only in medical writings, but in popular works, herbals and in art. The doctrine of diagnosis by the physical appearance of urine--color, density, sediments--stems from humoral physiology.

925. Hamarneh, Sami. "Arabic Medicine and its Impact on Teaching and Practice of the Healing Arts in the West." Oriente e occidente nel medioevo: Filosofia e scienze. Rome: Accademia nazionale dei Lincei, 1971, pp. 395-424.

Argues that medicine and pharmacy, which achieved a high degree of excellence in Islam, were transmitted to the West starting in the eleventh century. Many of the best treatises in Arabic medicine, pharmacy and surgery circulated widely, and Arabic medicine enjoyed a prestige equal to that of the Greek authorities.

926. Hammond, E.A. "Physicians in Medieval English Religious Houses." Bulletin of the History of Medicine, 32 (1958), 105-120.

Discusses the physicians and medical practice at major English foundations from the post-conquest period to the twelfth-century prohibitions against the study, teaching and practice of medicine by religious.

927. Hammond, E.A. "Incomes of Medieval English Doctors." Journal of the History of Medicine and Allied Sciences, 15 (1960), 154-169.

Explores the financial arrangements between physicians and patients during the latter part of the Angevin period. Fees appear to be related to the social standing of the physician and his reputation.

928. Hartley, Percival Horton-Smith and H.R. Aldridge. Johannes de Mirfield of St. Bartholomew's Smithfield, his Life and Works. Cambridge: Cambridge University Press, 1936. Pp. xiii + 191.

Provides the text and translation of extracts from Mirfield's Brevarium Bartholomei, as well as the text with translation of John's medical chapter from his theological treatise, the Florarium Bartholomei.

929. Isaac Israeli (Isḥāq al-Isrā'īlī). "Isaac Israeli Liber de definicionibus." Edited by J.T. Muckle. Archives d'histoire doctrinale et littéraire du moyen âge, 11 (1937-38), 299-340.

930. Jetter, Dieter. "Klosterhospitäler: St. Gallen, Cluny, Escorial." Sudhoffs Archiv: Zeitschrift für Wissenschaftsgeschichte, 62 (1978), 313-338.

Points out that medieval monasteries possessed more than one hospital, and that the office of hospitalarius was not identical with that of infirmarius.

931. Joannitius (Hunain ibn Isḥāq). The Book of the Ten Treatises on the Eye Ascribed to Hunain ibn Isḥāq (800-877 A. D.). Edited and translated by Max Meyerhof. Cairo: Government Press, 1928.

Arabic text and English translation. This work had been translated into Latin by Constantine the African.

932. Kallinich, Günter and Karin Figala. "Konrad von Eichstätt, eine Arztpersönlichkeit des Deutschen Mittelalters." Sudhoffs Archiv: Vierteljahrsschrift für Geschichte der Medizin und der Naturwissenschaften, der Pharmazie und der Mathematik, 52 (1968), 341-346.

Discusses the life and work of this German physician who wrote two medical works, a Regimen sanitatis and a Tractatus de qualitatibus ciborum.

933. Kallinich, Günter and Karin Figala. "Das Regimen sanitatis des Arnold von Bamberg." Sudhoffs Archiv: Zeitschrift für Wissenschaftsgeschichte, 56 (1972), 44-60.

Analyzes the content of this Regimen by the late thirteenth-early fourteenth-century prior and physician, finding that the work is anchored in the medieval medical tradition, although it does show a certain independence.

934. Kealey, Edward J. Medieval Medicus: A Social History of Anglo-Norman Medicine. Baltimore: Johns Hopkins University Press, 1981. Pp. x + 211.

Reveals the high level of health care in England during the first half of the twelfth century, discusses the careers of doctors who can be identified and the establishment of institutional facilities. Appendixes give a directory of physicians and a register of hospitals.

935. Kibre, Paul. "Hippocratic Writings in the Middle Ages." Bulletin of the History of Medicine, 18 (1946), 371-412.

Argues that Hippocratic writings had a continuous influence during the entire medieval period and shows the extent to which these works, whether authentic or spurious, dealing with general pathology, humoral theory, a practical regimen for health as well as the association between astrology and medicine, played a role in medical thought and study, as revealed by extant MSS, library catalogues, university statutes and early printed editions.

936. Kibre, Paul. "Dominicus de Ragusa, Bolognese Doctor of Arts and Medicine." Bulletin of the History of Medicine, 45 (1971), 383-386.

Presents further information on Dominicus, rectifying previously established key dates in his career and information regarding his medical writings.

937. Kibre, Pearl and Nancy G. Siraisi. "Matheolus of Perugia's Commentary on the Preface to the Aphorisms of Hippocrates." Bulletin of the History of Medicine, 49 (1975), 405-428.

Provides the text and translation of this commentary on the most popular of Hippocratic writings by a professor at Padua in the second half of the fifteenth century.

938. Kristeller, Paul O. "Philosophy and Medicine in Medieval and Renaissance Italy." Organism, Medicine, and Metaphysics: Essays in Honor of Hans Jonas. Edited by Stuart F. Spicker. Dordrecht:Reidel, 1978, pp. 29-40.

Shows that the connection between philosophy and medicine, not characteristic of the modern period, was especially manifest during the Middle Ages and the Renaissance, particularly at Bologna and other Italian universities where arts and medicine were subsumed under one faculty.

939. Kühnel, Harry. Mittelalterliche Heilkunde in Wien. Studien zur Geschichte der Universität Wien, 5. Graz: Böhlaus, 1965. Pp. 114.

Discusses clerical and lay physicians in the Viennese community before and after the establishment of the university and its medical faculty. Gives information on the careers of prominent medical masters at Vienna, personal physicians at the court at Vienna and surgeons.

940. Lauer, Hans Hugo. "Zur Beurteilung des Arabismus in der Medizin des mittelalterlichen England." Sudhoffs Archiv: Vierteljahrsschrift für Geschichte der Medizin und der Naturwissenschaften, der Pharmazie und der Mathematik, 51 (1967), 326-348.

Argues that Anglo-Saxon England had contacts with the east as shown by familiarity with oriental drugs. After the Norman conquest of England and Sicily, Salernitan materials were brought to England, and England fully participated in the reception of Arabic medicine during the period of translation. While early sources are sparse concerning medicine at English universities, early fourteenth-century lists of texts as Oxford reveal the assimilation of Arabic medicine.

941. Laux, Rudolf. "Ars medicinae: Ein frühmittelalterliches Kompendium der Medizin." Kyklos, 3 (1930), 417-434.

Provides the text and a summary of the contents of a collection of medical tracts in a Copenhagen MS of the eleventh century. Individual tracts from this MS can be found in other MSS, and it is assumed that they all derive from a common source. The content of the MS, especially its mythological introduction, is very similar to the Lectiones Heliodori.

942. Lemay, Helen R. "Arabic Influence on Medieval Attitudes toward Infancy." Clio Medica, 13 (1978), 1-12.

Argues that medieval authors before the translation of Arabic materials considered infancy as a sub-human period of life. With the reception of Arabic astrological and medical works, a new attitude was transmitted to the Latin West which proved influential. Arabic astrological works, which discussed the effect of the stars on the fetus and infant, considered early childhood an important stage in life. Arabic medical treatises likewise discussed the diseases of infancy, feeding and child care.

943. Lockwood, Dean Putnam. "Ugo Benzi: Medieval Philosopher and Physician, 1376-1439. Chicago: University of Chicago Press, 1951. Pp. xvi + 441.

Treats the career and works of Ugo Benzi of Siena who studied at Bologna and taught at several Italian universities. He wrote two philosophical works: a treatise on logic and a commentary on the Parva naturalia, as well as five great commentaries on Greek and Arabic medical authors. Focusses on Ugo's consilia, records of specific cases, deriving information from them as to Ugo's role as a physician and his relations with patients and colleagues.

944. Maccagni, Carlo. "Frammento di un codice di medicina del secolo XIV (manoscritto N. 735, già codice Roncioni N. 99) della Biblioteca Universitaria di Pisa." Physis, 11 (1969), 311-378.

Analyzes a MS at the University of Pisa dealing with cauterization and compares it to Ashmole MS 399 of the Bodleian and MS 49 of the Wellcome Institute of the History of Medicine.

945. MacKinney, Loren C. "Medieval Medical Dictionaries and Glossaries." Medieval and Historiographical Essays in Honor of James Westfall Thompson. Edited by James L. Cate and Eugene N. Anderson. Chicago: University of Chicago Press, 1938, pp. 240-268.

Discusses types of western European medical compendia in which information was alphabetically organized. These included general medical dictionaries, pharmaceutical works on simple and compound substances and glossaries of foreign and technical terms.

946. MacKinney, Loren C. "Medical Ethics and Etiquette in the Early Middle Ages: The Persistence of Hippocratic Ideals." Bulletin of the History of Medicine, 26 (1952), 1-31.

Argues that the high ideals of Hippocratic medicine prevailed in the early Middle Ages (400-1100 A. D.), as an examination of monastic and other writings reveals. Early medieval physicians reflect a blend of Hippocratic and Christian ethical concepts.

946a. MacKinney, Loren C. Medical Illustrations in Medieval Manuscripts. Berkeley: University of California Press; London: Wellcome Historical Medical Library, 1965. Pp. xvii + 263.

Presents (in Part I) illustrations from MSS showing the practice of medicine, surgery, orthopedics, obstetrics, dentistry, bathing and veterinary medicine with historical and explanatory introductions. Part II contains a check list of medieval miniatures in exant MSS compiled with the assistance of Thomas Heardon.

947. MacKinney, Loren and Harry Bober. "A Thirteenth-Century Medical Case History in Miniatures." Speculum, 35 (1960), 251-259.

Interprets a series of 8 illustrations without captions taken from a thirteenth-century MS as presenting a moral lesson: A lady who falls ill and rejects her physician's medicine, dies and is autopsied. Women then flock to the physician who cures them. The message is that one should heed the physician's advice or the result may be fatal. Bober has supplied a stylistic analysis of the miniatures.

948. McVaugh, Michael R. "Arnald of Villanova and Bradwardine's Law." Isis, 58 (1967), 56-64.

Argues that Bradwardine's widely influential exponential rule was an adaptation of an idea known among medical theorists at Montpellier. A geometrical increase in a medicinal quality had been related to an arithmetic increase in efficacy in the medical work of Alkindi, familiar to Arnald, whose pharmacy reflects Alkindian notions. Arnald's quantitative pharmacy may have reached Bradwardine through the latter's associate, Simon Bredon. Arnald himself was prepared to extend the application of this exponential relationship to quantitative descriptions of all dynamic processes.

949. McVaugh, Michael R. "Quantified Medical Theory and Practice at Fourteenth-Century Montpellier." Bulletin of the History of Medicine, 43 (1969), 397-413.

Argues that the application of quantitative relationships to the compounding of medicines from simples or herbs, introduced around 1300 at Montpellier by Arnald of Villanova, might have afforded an opportunity for the experimental verification of medical theory. But this potential was not fulfilled by Arnald or his successors. Arnald's quantitative pharmacy was of more value in natural philosophy than in medicine.

950. McVaugh, Michael R. "The Experimenta of Arnald of Villanova." Journal of Medieval and Renaissance Studies, 1 (1971), 107-118.

Examines this genuine work by Arnald, a collection of 73 experimenta--accounts of medical problems and types of treatment, composed during 1305-11. The work provides a glimpse of medical practice as carried on by Arnald and Montpellier physicians at the time when medicine at the university was developing a scholastic orientation, a change in which Arnald was a main figure. The *experimenta* reveal a surprisingly small range of medicines, and the courses of treatment are dissimilar to those suggested in Arnald's more professional and formal treatises; there is little scholastic medical theory. Includes a selection of *experimenta*.

951. McVaugh, Michael R. "The *humidum radicale* in Thirteenth-Century Medicine." *Traditio*, 30 (1974), 259-283.

Traces the advent of the radical moisture concept in the Latin West through the works of Galen, Isaac Israeli, Constantine and Avicenna, and the development of this idea by two Montpellier physicians, Bernard of Gordon and Arnald of Villanova.

952. McVaugh, Michael R. "An Early Discussion of Medicinal Degrees at Montpellier by Henry of Winchester." *Bulletin of the History of Medicine*, 49 (1975), 57-71.

Argues that the discussion of medical degrees in the commentary on the *Isagoge* of Joannitius (Hunain ibn Isḥāq), an early work by Henry of Winchester, chancellor of the medical faculty at Montpellier, 1239-40, offers a glimpse of medical writing at the university before the 1280's. Henry's use of medical degrees, the notion that the qualitative character of a substance can be measured on a scale of four degrees, predated the mathematical theory of compound medicines of Arnald of Villanova.

953. McVaugh, Michael R. "Two Montpellier Recipe Collections." *Science, Medicine and the University: 1200-1550. Essays in Honor of Pearl Kibre*, II. Edited by Nancy G. Siraisi and Luke Demaitre. *Manuscripta*, 20 (1976), pp. 175-180.

Describes two collections of medical recipes and treatments made at Montpellier around the mid-1330's and attributed to several physicians who taught at Montpellier, 1250-1345. Such collections, taken from many sources, were a common practice at Montpellier, and represent the result of practical experience and personal contact among physicians in the medical community at the university.

954. Mercier, Charles A. Leper Houses and Mediaeval Hospitals. London: H.K. Lewis, 1915. Pp. 48.

Traces the origin of leper houses, their nature as ecclesiastical foundations, their connections with hospitals and almshouses, their administration and provisions for the care of the ill.

955. Moonan, L. "Presurgical Sedation, Montpellier c. 1393: Testimony of Lawrence of Lindores." Medical History, 12 (1968), 299-301.

Offers evidence that sedatives were used at Montpellier at the end of the fourteenth century, as attested in the De anima commentary of Lawrence of Lindores.

956. O'Neill, Ynez Violé. "Michael Scot and Mary of Bologna: A Medieval Gynecological Puzzle." Clio Medica, 8 (1973), 87-111.

Argues that the account of the stones produced by a Mary of Bologna reported by Scot in three MSS of his translation of the De animalibus leads not to a diagnosis of calcified fibroid tumors, as previously thought, but rather the stones are evidence of a "missed abortion" of twins which degenerated and calcified, and was eventually expelled from the uterus.

957. Paniagua, Juan A. Estudios y notas sobre Arnau de Vilanova. Madrid: Consejo superior de investigaciones científicas, 1963. Pp. 82.

A collection of short notes including bibliographical information concerning Arnald's medical writings, the text of Arnald's small treatise on the diagnosis of leprosy, and bibliographical material pertaining to the alchemical treatises attributed to him.

958. Paniagua, Juan A. "Arnau de Vilanova, médico escolastico." Asclepio, 18 (1966-67), 517-532.

Argues from an analysis of the genuine Arnaldian corpus that the Montpellier physician was a thoroughgoing scholastic with a reverence for Galen. His criticism of Galen was confined to a reinterpretation of the Greek master and his allegiance to other authorities depended on their adherence to Galen.

959. Paniagua, Juan A. El maestro Arnau de Vilanova, médico. Valencia: Catedra e Instituto de Historia de la Medicina, 1969. Pp. 92.

Surveys Arnald's life and career with particular emphasis on his medical works and commentaries on ancient medical authors written at Montpellier and elsewhere. Likewise discusses Arnald's treatises on pharmacology, magic and astrology, and the alchemical books attributed to him.

960. Pinto, Lucille B. "The Folk Practice of Gynecology and Obstetrics in the Middle Ages." Bulletin of the History of Medicine, 47 (1973), 513-523.

Describes a MS from the end of the eleventh or beginning of twelfth century, now at Bern, containing medical recipies, but also dealing with medical problems of women and with childbirth. The MS is of monastic origin, and is a blend of practical treatment with folk tradition.

961. Pinto, Lucille B. "Medical Science and Superstition: A Report on a Unique Medical Scroll of the Eleventh-Twelfth Century." Manuscripta, 17 (1973), 12-21.

Reports on the content of a medical roll in the Bürgerbibliothek at Bern, dating from the end of the eleventh century and the beginning of the twelfth, of clerical origin. Written in German and Latin, it describes the medical use of various plants and animals intended for humans and domestic animals; there are also items which employ magic. The work is a blend of medical knowledge and folk superstition which, in Germany at least, had a long career.

962. Poulle-Drieux, Yvonne. "L'hippiatrie dans l'Occident latin du XIIIe au XVe siècle." Médecine humaine et vétérinaire à la fin du moyen âge. Edited by Guy Beaujouan. Paris: Minard, 1966, pp. 11-167.

Provides a sketch of several texts on hippiatry plus a discussion of pathology, medical and surgical treatment and materia medica. Points out that medieval hippiatry drew on classical sources without Arabic intermediaries, that most texts stem from the Mediterranean region and that astrology played a minor role in dealing with diseases of horses.

963. Rhazes (al-Rāzi). A Treatise on the Smallpox and Measles. Translated by William A. Greenhill. London, 1848. Reprint. Baltimore: Williams & Wilkins, 1939. Pp. 84.

964. Riddle, John M. "Theory and Practice in Medieval Medicine." Viator, 5 (1974), 157-184.

Argues from the perspective of drug therapies that early medieval medicine was a practical skill without a conceptual framework which, while dependent on a legacy of older knowledge, was open to innovative additions from experience and folk medicine. The recovery of ancient texts and the transmission of Arabic treatises together with developments in medieval education, particularly in the university, created a professional class of physicians who wrote and taught a scholastic theoretical medicine removed from the requirements of ordinary practice and often not apt for the treatment of patients. The gap between theoretical medicine and practice widened; western medicine, however, was saved from complete domination by theory thanks to the anti-intellectual currents of the late medieval period.

965. Rosen, Edward. "The Invention of Eyeglasses." Journal of the History of Medicine and Allied Sciences, 11 (1956), 13-46; 183-218.

Discusses in detail the difficult problem of determining the medieval inventor of eyeglasses, an innovation attributed to several persons in a variety of localities.

966. Rubin, Stanley. Medieval English Medicine. New York: Barnes & Noble, 1974. Pp. 232.

Surveys medical thought from the Anglo-Saxon period to the beginning of the fourteenth century. Discusses the nature of the Anglo-Saxon leech, his training and types of treatment. Focuses on the problem of leprosy, monastic infirmaries and secular hospitals, and the transmission of Arabic and Greek medical authors as seen in the work of Bartolomeus Anglicus, Gilbertus Anglicus and John of Gaddesden.

967. Schipperges, Heinrich. "Zur Unterscheidung des Physicus vom Medicus bei Petrus Hispanus." Asclepio, 22 (1970), 321-327.

Argues that the competition between the titles physicus and medicus ceased in the mid-twelfth century with the increased prestige of the appellation physicus, thanks to the recovery of Aristotle and the reception of Arabic scientific medicine. Henceforth a distinction was drawn between the physicus who, as a philosopher, was concerned with general laws and principles, and the medicus, who as an empirical observer, was concerned with a particular human situation. This development can be seen in the works of Peter of Spain.

968. Schipperges, Heinrich. Arabische Medizin im lateinischen Mittelalter. Sitzungsberichte der Heidelberger Akademie der Wissenschaften, math.-naturwiss. Klasse, 2. Berlin: Springer, 1976. Pp. 192.

Surveys the transmission and absorption of Arabic medical traditions into the Latin West as divided into three stages: The reception of Arabic science during the eleventh and twelfth centuries, a period of assimilation, and lastly, the full integration of this new medical knowledge during the thirteenth and fourteenth centuries, a development directly connected with the university.

969. Seyfert, Werner. "Ein Komplexionentext einer Leipziger Incunabel (angeblich eines Johann von Neuhaus) und seine handschriftliche Herleitung aus der Zeit nach 1300." Archiv für Geschichte der Medizin, 20 (1928), 272-299, 372-389. Reprint. Wiesbaden: Steiner, 1965.

Discusses the doctrine of the complexio, the division of individuals into physiological and psychological types of temperaments according to the four elements, their qualities and the four humors. Describes a text printed at Leipzig about 1500 and attributed to a Johannes de Novo Domo (actually John of Paris; text, pp. 286-299).

970. Singer, Charles, ed. The Fasciculo di medicina, Venice, 1493. 2 vols. Florence: R. Lier, 1925. Pp. 112 + [71].

A facsimile of the medieval collection of medical treatises associated with Johannes de Ketham in the Italian version of 1493. The largest item is the Anatomy of Mondino de'Luzzi (1316); provides a translation of the latter.

971. Siraisi, Nancy G. "The Music of Pulse in the Writings of Italian Academic Physicians (Fourteenth and Fifteenth Centuries)." Speculum, 50 (1975), 689-710.

Maintains that the connection between music and the beat of the pulse was of interest to physicians at northern Italian universities, many of whom accepted pulse music to some degree. Their discussions reflect a long tradition of a science of pulse; their authority on music was Boethius.

972. Siraisi, Nancy G. "Taddeo Alderotti and Bartolomeo da Varignana on the Nature of Medical Learning." Isis, 68 (1977), 27-39.

Argues that these two masters of the faculty of arts and medicine at Bologna differed on the nature of medical learning. Taddeo stressed traditional connections between the arts, philosophy and medicine and relied strongly on Aristotle. Bartolomeo, his pupil and colleague, criticized the link between medicine and philosophy, refusing either to subordinate medicine to natural science or to view it as a subdivision of natural philosophy. See item 974.

973. Siraisi, Nancy G. "The Medical Learning of Albertus Magnus." Albertus Magnus and the Sciences: Commemorative Essays. Edited by James A. Weisheipl. Toronto: Pontifical Institute of Mediaeval Studies, 1980, pp. 379-404.

Surveys Albert's view of medical learning and the works on medicine used by him. Concludes that Albert held that the natural philosophy of Aristotle could be integrated with medical knowledge.

974. Siraisi, Nancy G. Taddeo Alderotti and his Pupils: Two Generations of Italian Medical Learning. Princeton: Princeton University Press, 1981. Pp. xxiii + 461.

Surveys the works, careers and medical activities of Taddeo, whose academic career was spent at Bologna, and of his students and/or colleagues. Emphasizes their views of the nature of medical learning and medicine's position vis-à-vis the other sciences, their attitudes toward the use of natural philosophy and scholastic questions in teaching, as well as the knowledge of Galen revealed in their works and the relationship between theory and practice shown in their work and writings.

975. Sudhoff, Karl. "Die pseudo-hippokratische Krankheitsprognostik nach dem Auftreten von Hautausschlägen, Secreta Hippocratis oder Capsula eburna benannt." Archiv für Geschichte der Medizin, 9 (1915-16), 79-116. Reprint. Wiesbaden: Steiner, 1964.

Traces the origin of this work on prognostications according to the appearance of skin eruptions. Stemming from the Mediterranean area in the fifth-sixth centuries, it was first translated into Latin between the sixth-eighth centuries. Translated into Arabic, it was then translated by Gerard of Cremona as the Liber veritatis. Gives the text from an undated printing of the 1580's.

976. Sudhoff, Karl. "Pestschriften aus dem 150 Jahren nach der Epidemie des 'schwarzen Todes' 1348." Archiv für Geschichte der Medizin, 4 (1910-11), 191-222, 389-424; 5 (1911-12), 36-87, 332-396; 6 (1912-13), 313-379; 7 (1913-14), 57-114; 8 (1914-15), 175-215, 236-289; 9 (1915-16), 53-78, 117-167; 11 (1918-19), 44-92, 121-176; 14 (1923), 1-25, 79-105, 129-168; 16 (1924-25), 1-76, 77-188; 17 (1925), 12-139, 241-291. Reprint. Wiesbaden: Steiner, 1964-65.

Texts of European plague tracts.

977. Tacuinum sanitatis. Das Buch der Gesundheit. Edited by Luisa Cogliati Arano, with an introduction by Heinrich Schipperges and Wolfram Schmitt. Munich: Heimeran, 1976. Pp. 153.

The second part of the Tacuinum dei, translated from Arabic in the thirteenth century.

978. Talbot, Charles H. "The Accumulationes receptarum of Joannes Piscis." Bulletin of the History of Medicine, 34 (1960), 123-136.

Argues that a study of the contents of this work by Piscis, master at Montpellier and chancellor of the university from 1394-1433, a collection of recipes in sixty-six chapters, reveals little progress in medical knowledge over that of the ninth century. It aids in understanding the decline of the Montpellier faculty of medicine in the fifteenth century.

979. Talbot, Charles H. "A Mediaeval Physician's Vade mecum." Journal of the History of Medicine and Allied Sciences, 16 (1961), 213-233.

Describes a MS consisting of small sheets of parchment which were folded so as to be carried in a bag at the belt, containing a calendar, astronomical-astrological tables, rules for bloodletting and descriptions of urines, which would have enabled the physician to make a diagnosis. Text and translation.

980. Talbot, Charles H. "Some Notes on Anglo-Saxon Medicine." Medical History, 9 (1965), 156-169.

Argues that the excellence of Salernitan medicine compared to Anglo-Saxon is not confirmed by the evidence. Shows that the same texts that were used at Salerno were also known in England two centuries before this transmission was supposed to have occurred.

981. Talbot, Charles H. Medicine in Medieval England. London: Oldbourne, 1967. Pp. 222.

Traces medical thought from the Anglo-Saxon leechbooks to the appearance of medical works in English in the late fourteenth and early fifteenth centuries.

982. Talbot, Charles H. "Dame Trot and her Progeny." Essays and Studies, 25 (1972), 1-14.

Discusses women doctors in the Middle Ages.

983. Thomasen, Anne-Liese. "Historia animalium contra Gynaeciae in der Literatur des Mittelalters." Clio Medica, 15 (1980), 5-23.

Contrasts the dominance of Aristotle's biological works over the medical tradition stemming from Alexandria which appeared in the medieval West with translations of Soranus in literature regarding embryology, gynecology and diseases of women. While a few medieval doctors attempted to counter the Aristotelian tradition, it was not until the end of the Middle Ages that Soranus' views were of influence.

984. Thorndike, Lynn. "Medieval Sanitation, Public Baths and Street Cleaning." Speculum, 3 (1928), 192-203.

Maintains that the supposed lack of medieval concern for personal hygiene and public health has been exaggerated. Cities had municipal physicians, and there were many hospitals as well as charitable institutions for care of the needy ill. An awareness of the contagion of disease led to quarantine regulations, espcially after the Black Death. City ordinances attempted to control the state of streets, while bathing was stressed in medical works; public baths were commonplace.

985. Thorndike, Lynn. "De complexionibus." Isis, 49 (1958), 398-408.

Presents several MS treatments of the doctrine of the complexio, the physical constitution of the body as a whole as determined by the four humors.

986. Torre, Esteban. Averroes y la ciencia medica: La doctrina anatomofucional del Colliget. Madrid: Ediciones del Centro, 1974. Pp. 276.

Presents an analysis of the content of the medical encyclopedia, the Colliget, with particular emphasis on the relationship between form (anatomy) and function. Concludes that Averroes believed that medicine was to be founded on principles of natural philosophy. Points out tha primary role of deductive method in his view of medicine and the absolute predominance in his anatomy and physiology of function over form in which human organs adopt certain forms in response to operations arising from nature. Gives translations of the Introduction and Books I and II of the Colliget.

987. Welborn, M.C. "The Long Tradition: A Study in Fourteenth-Century Medical Deontology." Medieval and

Historiographical Essays in Honor of James Westfall Thompson. Edited by James L. Cate and Eugene N. Anderson. Chicago: University of Chicago Press, 1938, pp. 240-268.

Argues that medieval physicians had high ideals of professional conduct which stressed ethical and humane attitudes toward their patients.

988. Wickersheimer, Ernest. "Faits cliniques observés à Strasbourg et à Haslach en 1362 et suivis de formules de remèdes." Bulletin de la société française d'histoire de la médicine, 33 (1939), 69-92.

Describes and gives the text of a medical work, possibly by Tilliman de Syberg. The first part contains clinical observations made by the author himself; the second part provides remedies recommended by the author, including veterinary material.

989. Wood, Charles. "The Doctor's Dilemma: Sin, Salvation and the Menstrual Cycle in Medieval Thought." Speculum, 56 (1981), 710-727.

Argues that medieval views of menstruation were conditioned by a theological misogyny which saw this physiological phenomenon as the "curse of Eve," re-enforced by classical opinion as to the purpose of the menses. The theological aspects were particularly troubling in the case of the Virgin.

990. Woodings, Ann F. "The Medical Resources and Practice of the Crusader States in Syria and Palestine." Medical History, 15 (1971), 268-277.

Points out that the Crusaders found excellent native physicians in Outremer, as well as well-established facilities for the care of the sick and for medical teaching. The Crusaders took advantage of these opportunities and built on them, as can be seen in the medical work of the Military Orders, especially that of St. John.

991. Wright, C.E., ed. Bald's Leechbook. British Museum Royal MS 12. D. xvii. Introduction by Bertram Colgrave and appendix by Ralph Quirk. Baltimore: Johns Hopkins Press, 1955. Pp. 258.

992. Ziegler, Philip. The Black Death. New York: John Day Co., 1969. Pp. 312.

An account of the Black Death with an extensive bibliography.

## ANATOMY AND PHYSIOLOGY

993. Alfred of Sareshel. Des Alfred von Sareshel (Alfredus Anglicus) Schrift De motu cordis. Edited by Clemens Baeumker. Beiträge zur Geschichte der Philosophie des Mittelalters, 23. Münster: Aschendorff, 1923. Pp. xviii + 114.

Edition of the text.

994. Aquinas, Thomas. "St. Thomas Aquinas on the Movement of the Heart." Translated by Vincent R. Larkin. Journal of the History of Medicine and Allied Sciences, 15 (1960), 22-30.

Provides the translation and outline of a letter, De motu cordis, addressed to a Master Philip.

995. Baader, Gerhard. "Zur Anatomie in Paris im 13. und 14. Jahrhundert." Medizinhistorisches Journal, 3 (1968), 40-53.

Discusses the anatomical works available at Paris during this period, their authors and their provenance.

996. Demaitre, Luke E. and Anthony A. Travill. "Human Embryology and Development in the Works of Albertus Magnus." Albertus Magnus and the Sciences: Commemorative Essays. Edited by James A. Weisheipl. Toronto: Pontifical Institute of Mediaeval Studies, 1980, pp. 403-440.

Argues that theological and medical traditions were less important in Albert's treatment of embryology than was the natural philosophy derived from Aristotle. Turning to several main embryological problems in his De animalibus, concludes that Albert produced a complete coverage of the knowledge available, combining Aristotle, hear-say and personal observation.

997. French, Roger. "A Note on the Anatomical Accessus of the Middle Ages." Medical History, 23 (1979), 461-468.

Argues that the anatomical version of the medical accessus, a formal introduction to a work giving the topics to be covered, reached its height in the same period when human dissection was first used in medical teaching.

998. Hewson, M. Anthony. Giles of Rome and the Medieval Theory of Conception: A Study of the De formatione corporis humani in utero. London: Athlone Press, 1975. Pp. viii + 268.

Outlines the principal arguments of Giles' treatise on embryology and evaluates its position vis-à-vis contemporary philosophical and theological issues. Written around 1274-78, the work likewise reflects the interest in generation sparked by the biological writings of Aristotle, Avicenna and Averroes. Giles rejected the Galenic notion that the female plays an active role in generation; he attempted to harmonize medical and philosophical views. His treatise aroused comment from subsequent medical writers at Bologna who, although representing a more "scientific" tradition, regarded Giles' work with esteem.

999. Hill, Jr., Boyd H. "Another Member of the Sudhoff Fünfbilderserie--Wellcome MS. 5000." Sudhoffs Archiv für Geschichte der Medizin und der Naturwissenschaften, 43 (1959), 13-19.

Describes this version of the five picture series, giving the text accompanying it. See item 1003.

1000. Hill, Jr., Boyd H. "The Grain and the Spirit in Mediaeval Anatomy." Speculum, 40 (1965), 63-73.

Draws attention to the notion that the arteries originate in a "black grain" in the heart in close connection with the spiritus, also in the heart, a factor at times given a physical meaning identified with breath, but also viewed as of spiritual or theological significance mediating between body and soul. The black grain can be clearly seen in some anatomical drawings.

1001. Kurdziałek, Marian. "Anatomische und embryologische Äusserungen Davids von Dinant." Sudhoffs Archiv für Geschichte der Medizin und der Naturwissenschaften, 45 (1961), 1-22.

Gives the text of an anatomical and embryological fragment of David of Dinant.

1002. MacKinney, Loren C. "The Beginnings of Western Scientific Anatomy: New Evidence and a Revision in Interpretation of Mondeville's Role." Medical History, 6 (1962), 233-239.

Argues that the development of anatomy was a gradual evolution more dependent on anatomical illustrations than on dissections. Focuses on a revaluation of Mondeville, arguing that his illustrations of 1304 at Montpellier reveal little advance over the thirteenth century while the drawings from his Paris lectures were not particularly influential. However, there was, in any event, a trend toward increasing naturalism in anatomical drawings from the twelfth to the fifteenth centuries.

1003. MacKinney, Loren C. and Boyd H. Hill, Jr. "A New Fünfbilderserie Manuscript--Vatican Palat. Lat. 1110." Sudhoffs Archiv für Geschichte der Medizin und der Naturwissenschaften, 48 (1964), 323-330.

Reports and describes another version of the group of anatomical illustrations first published by Sudhoff. See item 999 and Luigi Belloni, "Gli schemi anatomici trecenteschi ... del codice Trivulziano 836," Rivista di storia delle scienze, med. e nat., 40 (1950), 193-207.

1004. O'Neill, Ynez Violé. "The Fünfbilderserie Reconsidered." Bulletin of the History of Medicine, 43 (1969), 236-245.

Argues that this series of anatomical diagrams published by Sudhoff ["Anatomische Zeichnung (Schemata) aus dem 12. und 13. Jahrhundert und eine Skelettzeichnung des 14. Jahrhunderts." Studien zur Geschichte der Medizin, I. Leipzig: Barth, 1907], which purportedly describes the systems of the body according to Galen, contains a distortion of Galen's ideas on the arterial system.

1005. O'Neill, Ynez Violé. "Innocent III and the Evolution of Anatomy." Medical History, 20 (1976), 429-433.

Maintains that the decretals of Innocent contain evidence that the first public dissections in Christian Europe were postmortem examinations for legal purposes.

1006. Shaw, James R. "Scientific Empiricism in the Middle Ages: Albertus Magnus on Sexual Anatomy and Physiology." Clio Medica, 10 (1975), 53-64.

Argues that Albert's strong belief in an empirical approach to nature in which he not only differed from his authorities, but added observations of his own, can be seen in the treatment of sexual anatomy in his commentary on the De animalibus.

1007. Sudhoff, Karl. Ein Beitrag zur Geschichte der Anatomie im Mittelalter, speziell der anatomischen Graphik, nach Handschriften des 9. bis 15. Jahrhunderts. Studien zur Geschichte der Medizin, 4. Leipzig, 1908. Reprint. Hildesheim: Olms, 1964. Pp. v + 94 + [46].

A selection of 7 collections of anatomical drawings and their texts.

1008. Sudhoff, Karl. "Weitere Beiträge zur Geschichte der Anatomie im Mittelalter." Archiv für Geschichte der Medizin, 7 (1913-14), 363-378; 8 (1914-15), 1-21, 129-141; 10 (1916-17), 255-263. Reprint. Wiesbaden: Steiner, 1964.

Further examples of anatomical MSS with illustrations.

1009. Sudhoff, Karl. "Richard der Engländer." Janus, 28 (1924), 397-403.

Discusses the identity and the works of the physician and medical author, Richard Anglicus. Richard was connected with Paris and Montpellier; there is no reason to associate him or his Anatomy with Salerno. Gives the text of the prologue to Richard's Micrologus. See item 1010.

1010. Sudhoff, Karl. "Der Micrologus-Text der Anatomia Richards des Engländers." Archiv für Geschichte der

Medizin, 19 (1927), 209-239. Reprint. Wiesbaden: Steiner, 1965.

Provides the text of the Anatomia of Richard Anglicus, part of his compendium, the Micrologus.

1011. Wickersheimer, Ernest. "L'Anatomie de Guido da Vigevano, médecin de la reine Jeanne de Bourgogne (1345)." Archiv für Geschichte der Medizin, 7 (1913-14), 1-25. Reprint. Wiesbaden: Steiner, 1964.

Describes the content and gives the text of Guido's anatomical work. Divided into four parts (introduction, veins and blood, explanations of the figures, swelling of the uvula and palate), the treatise is profusely illustrated, giving evidence of Guido's experience with dissection in Italy. See item 1012.

1012. Wickersheimer, Ernest. Anatomies de Mondino dei Luzzi et de Guido de Vigevano. Paris, 1926. Reprint. Geneva: Slatkine, 1977. Pp. 91.

Facsimile reprint of Mondino's Anatomy from the Pavia, 1478 edition, and the text of the Anothomia ... designata per figuras of de Vigevano, with notes.

1013. William of Saliceto. Die Anatomie des Wilhelm von Saliceto. Edited by F.O. Schaarschmidt. Leipzig: Noske, 1919. Pp. 75.

1014. William of Saliceto. L'Anatomia di Guglielmo da Saliceto. Commentary by Guiseppe Caturegli. Scientia veterum, 136. Pisa: Glardini, 1969. Pp. 115.

A facsimile of the text in a Lucca MS.

## SURGERY

1015. Albucasis (Abulcasis). Albucasis on Surgery and Instruments. Edition of the Arabic text with English translation and commentary by M.S. Spink and G.L. Lewis. Berkeley: University of California Press, 1973. Pp. xv + 850.

1016. Arderne, John. Treatises of Fistula in ano. Edited by D'Arcy Power. London: Kegan Paul, Trench, Trübner, 1910. Pp. xxxvii + 156.

1017. Arderne, John. De arte phisicali et De cirurgia. Translated by D'Arcy Power. London: J. Bale Sons & Danielson, 1922. Pp. vii + 60.

1018. Chauliac, Guy de. La Grande Chirurgie de Guy de Chauliac. Edited by Édouard Nicaise. Paris: Félix Alcan, 1890. Pp. cxci + 747.

* Hartley, Percival Horton-Smith and H.R. Aldridge. Johannes de Mirfield of St. Bartholomew's Smithfield, his Life and Works. Cambridge: Cambridge University Press, 1936. Pp. xiii + 191.

Cited as item 928.

1019. Huard, Pierre and Mirko Dražen Grmek. Mille ans de chirurgie en occident: Ve-XIe siècles. Paris: Roger Dacosta, 1966. Pp. 184.

Surveys the most productive period of medieval surgery beginning with the treatises of Constantine the African and the assimilation of Greek and Arabic medicine to the early sixteenth century. Discusses the treatises of Roger of Salerno, Theodoric, William of Saliceto, Henry of Mondeville and Guy de Chauliac.

1020. Mirfeld, John of. John of Mirfeld (d. 1407), Surgery: A Translation of his Breviarium Bartholomei, Part IX. Translated by James B. Colton with an introduction by Frank B. Berry. New York: Hafner, 1969. Pp. xix + 230.

1021. Mondani, Pier Luigi. Autori e fonti chirurgiche mediovali tratte dalla Chirurgia di Pietro Arzellata. Scientia veterum, 129. Pisa: Giardini, 1968, Pp. 128.

Gives the text and Italian translation of the introduction to Pietro's Surgery. Pietro studied at Montpellier and taught there and at Bologna. Also provides lists of the chapter titles of the work and the authors cited.

1022. Mondeville, Henry of. La Chirurgie de Henri de Mondeville. Edited by Édouard Nicaise. Paris: Félix Alcan, 1893. Pp. lxxxii + 902.

1023. Ogden, Margaret S. "Guy de Chauliac's Theory of the Humors." Journal of the History of Medicine and Allied Sciences, 24 (1969), 272-291.

Argues that Guy modified Galen's theory of the humors, making a distinction between natural and unnatural humors. These two classes were distinguished by two different criteria: nutritive and non-nutritive, normal and abnormal. His normative division differed from Galen and Avicenna; the division into nutritive and non-nutritive is dissimilar to anything in his sources.

1024. Ogdon, Margaret S. "The Galenic Works Cited in Guy de Chauliac's Chirurgia magna." Journal of the History of Medicine and Allied Sciences, 28 (1973), 24-33.

Lists 38 works of Galen in Guy's Surgery, identifying the version he used.

1025. Roland of Parma. Chirurgia Rogerii per Rolandum Parmensem (Codex Ambrosianus, I, 18). Edited by Antonius Nalesso. Rome: Università degli studi di Roma. 1968. Pp. 133.

1026. Sudhoff, Karl. Beiträge zur Geschichte der Chirurgie im Mittelalter. Studien zur Geschichte der Medizin, 10-12. Leipzig: Barth, 1914, 1918. Pp. x + 224, xvi + 685.

A two-part study of surgical materials from MS sources, particularly the miniatures. Part one deals with surgical procedures, wounds, cupping and bleeding, including illustrations which served for teaching purposes and as aids to the practicing surgeon. Part II contains illustrative material on surgical instruments as well as surveys of and selections from many surgical texts.

1027. Tabanelli, Mario. La chirurgia italiana nell'altro medioevo. 2 vols. Florence: Olschki, 1965. Pp. 1075.

Provides extensive selections in Italian translation of the surgeries of Roger and Roland of Parma,

Theodoric (bishop of Cervia), William of Saliceto and Lanfranc of Milan.

1028. Tabanelli, Mario. Un secolo d'oro della chirurgia francese (1300). I: Henry de Mondeville. II: Guy de Chauliac. 2 vols. Forlî: Valbonesi, 1969-70.

Italian translations of Henry of Mondeville's Chirurgia and Guy de Chauliac's Chirurgia magna.

1029. Tabanelli, Mario. Un chirurgo italiano del 1200: Bruno da Longoburga. Florence: Olschki, 1970. Pp. 146.

Provides a summary of the chapters of this thirteenth-century physician's Chirurgia magna (1252), with extensive translations of important passages. Bruno, who may have studied medicine at Bologna, also composed a Chirurgia parva, a brief compendium written after his Great Surgery.

1030. Tabanelli, Mario. Tecniche e strumenti chirurgici del XIII e XIV secolo. Florence: Olschki, 1973. Pp. ix + 174.

Provides chapter summaries of several major surgical treatises, including those of Albucasis, Roger and Roland of Parma, Bruno of Longoburgo, William of Saliceto, Lanfranc of Milan, Henry of Mondeville, Jean Yperman, John Arderne and Guy de Chauliac.

1031. Theodoric, Bishop of Cervia. The Surgery of Theodoric ca. A. D. 1267. 2 vols. Translated by Eldridge Campbell and James Colton. New York: Appleton-Century-Crofts, 1955, 1960. Pp. xl + 222, xix + 233.

English text only. Volume 1 contains Books I and II; volume 2, Books III and IV.

## PHARMACY AND MATERIA MEDICA

1032. Dillemann, G. "La pharmacopée au moyen âge." Revue d'histoire de la pharmacie, 19 (1968), 163-170.

Reviews briefly the scholarship on the major collections of recipes of herbal drugs.

1033. Forbes, Thomas R. "Medical Lore in the Bestiaries." _Medical History_, 12 (1968), 245-253.

Points out that curative properties were assigned to a number of bestiary animals; for example, the _caladrius_ bird, the beaver and the stag.

1034. Gagnière, S. "Les apothicaires à la cour des papes d'Avignon." _Revue d'histoire de la pharmacie_, 23 (1976), 145-157.

Exploits the Vatican archives for information concerning apothecaries who were official suppliers of drugs, spices and other substances to the papal court. Among these, one during each pontificate held the position of papal apothecary. Provides a biographical list of known apothecaries during the pontificates of John XXII, Benedict XII, Clement VI and Innocent VI.

1035. Granel, F. "La thériaque de Montpellier." _Revue d'histoire de la pharmacie_, 23 (1976), 75-83.

Traces the development and use of theriac at Montpellier.

1036. Levey, Martin. "Influence of Arabic Pharmacology on Medieval Europe." _Oriente e occidente nel medioevo: Filosofia e scienze_. Rome: Accademia nazionale dei Lincei, 1971, pp. 431-444.

Argues that Arabic influences on European pharmacology is well-attested. Among the Arabs, pharmacological knowledge transcended the Greek heritage and was enriched by information from China, India and Persia. Arabic treatises in translation and as the basis for compilations became a major part of the European tradition.

1037. McVaugh, Michael R. "_Apud antiquos_ and Mediaeval Pharmacology." _Medizinhistorisches Journal_, 1 (1966), 16-23.

Argues that the treatise _Apud antiquos_, edited here, is an example from a transitional period in pharmacology between the reception of classical theories explaining the behavior of simple medicines in terms of degrees introduced by Constantine, and the application of this theory to compound medicines found in the late thirteenth and fourteenth centuries.

1038. McVaugh, Michael R. "Theriac at Montpellier 1285-1325. (With an Edition of the Questiones de tyrica of William of Brescia). Sudhoffs Archiv: Zeitschrift für Wissenschaftsgeschichte, 50 (1972), 113-144.

Examines the medical and philosophical issues raised at Montpellier in connection with the compound medicine, theriac, famous since antiquity as a remedy against poisons, and relates these discussions to the shift in medicine at Montpellier from an empirical to a more theoretical and philosophical orientation.

1039. Marbode of Rennes. De lapidibus Considered as a Medical Treatise. Edited with commentary and with C.W. King's translation, and the text and translation of Marbode's minor works on stones by John M. Riddle. Sudhoffs Archiv: Zeitschrift für Wissenschaftsgeschichte. Beiheft 20. Wiesbaden: Steiner, 1977. Pp. xii + 144.

An edition of Marbode's (1095-1123) popular poem treating 60 stones. Argues from the work's association with herbals and medical MSS that it was considered as a guide to the medical use of stones. Marbode's chief source was the first century lapidary of Damigeron, but he drew from a wide variety of materials. The English translation is based on that made by C.W. King in 1860, corrected when necessary. Marbode also wrote three smaller lapidaries, one in verse.

1040. Maviglia, Alessandro. "Le prime tariffe italiane dei medicinali." Vorträge der Hauptversammlung der internationalen Gesellschaft für Geschichte der Pharmazie. Stuttgart: Wissenschaftliche Verlag, 1970, pp. 95-105.

Discusses the regulations of Frederick II, promulgated between 1231 and 1240, which contain the first rules for fixing drug prices in Italy.

1041. Medicina antiqua-libri quattuor medicinae. Codex Vindobonensis 93 der Österreichischen

*Nationalbibliothek*. 2 vols. Commentary volume with medico-historical introduction by Charles H. Talbot and with an introduction on the manuscript and iconography by Franz Unterkirchen. Graz: Akademische Druck- u. Verlagsanstalt, 1972.

Contains a facsimile reproduction of a medieval MS with four works on *materia medica*.

1042. Multhauf, Robert P. "John of Rupescissa and the Origin of Medical Chemistry." *Isis*, 45 (1954), 359-367.

Argues that the alchemical *Liber lucis* of this fourteenth-century Catalan apocalyptic preacher typifies a late medieval alchemy which linked alchemical theory with actual chemical procedures. The *De consideratione quintae essentiae* (probably genuine), in which alcohol is likened to the Aristotelian quintessence, with its "fifth essences" of many chemical substances, foreshadowed the medical chemistry of the sixteenth century.

1043. Riddle, John M. "Amber and Ambergris in Plague Remedies." *Sudhoffs Archiv für Geschichte der Medizin und der Naturwissenschaften*, 48 (1964), 111-122.

Points out that both amber and ambergris were used as aromatic substances to free the air from plague. The more powerful ambergris appears more often in *materia medica*, although the two substances were frequently confused; ambergris was expensive in comparison to the more available amber.

1044. Riddle, John M. "The Introduction and Use of Eastern Drugs in the Early Middle Ages." *Sudhoffs Archiv für Geschichte der Medizin und der Naturwissenschaften*, 49 (1965), 185-198.

Argues that an examination of pre-Salernitan recipe literature reveals a large number of eastern drugs, unknown in antiquity, and thus gives evidence of continuous contacts between western Europe and the East before the known transmission of medical texts during a period of supposed European isolation and Arabic domination of the eastern Mediterranean. The presence of such drugs is revealed in early medical antidotaries such as a ninth-century work at St. Gall, and in the ninth-century Anglo-Saxon Lacnunga.

1045. Riddle, John M. "Lithotheraphy in the Middle Ages." Pharmacy in History, 12 (1970), 39-50.

Argues that there is much evidence that stones and minerals were considered important in treatment by medieval physicians. A study of over 600 MSS reveals that lapidaries were part of the medical literature, often associated in the MSS with medical works and herbals. The relation between lapidaries and medicine began in the late eleventh century, represented but not initiated by the De lapidibus of Marbode of Rennes.

1046. Sigerist, Henry E. "Materia medica in the Middle Ages." Bulletin of the History of Medicine, 7 (1939), 417-421.

Traces the fortunes of the versions of the Materia medica of Dioscorides in the Latin West. A Latin edition of the alphabetical version may be a product of Salerno.

1047. Stannard, Jerry. "Greco-Roman Materia medica in Medieval Germany." Bulletin of the History of Medicine, 46 (1972), 455-468.

Establishes that the medical texts of antiquity, especially those of Galen and Hippocrates, provided the basis of a classical component in German writings on material medica.

1048. Stannard, Jerry. "Marcellus of Bordeaux and the Beginnings of Medieval materia medica." Pharmacy in History,15 (1973), 47-53.

Argues that materia medica stemming from antiquity was modified in the medieval period. This change can be seen in the collection of recipies by Marcellus (Liber de medicamentibus, A. D. 395-410).

1049. Voigts, Linda E. "Anglo-Saxon Plant Remedies and the Anglo-Saxons." Isis, 70 (1979), 250-268.

Argues that the large corpus of extant Anglo-Saxon information on medicinal plants and herbs does not represent mindless copying of materials, nor are the illustrations merely stylized drawings of plants that could not have been grown in England. Rather they reveal an intimate and practical acquaintance with living forms.

1050. Zimmermann, Volker. "Der Rosmarin als Heilpflanze und Wunderdrogen: Ein Beitrag zu den mittelalterlichen Drogenmonographien." Sudhoffs Archiv: Zeitschrift für Wissenschaftsgeschichte, 64 (1980), 351-370.

Discusses medieval works on the properties and uses of individual drugs, particularly rosemary, known for its medicinal properties since antiquity. Provides the text of a German work on rosemary.

## PSYCHOLOGY

1051. Aquinas, Thomas. In Aristotelis librum De anima commentarium. Edited by A.-M. Pirotta. Turin: Marietti, 1948. Pp. viii + 223.

1052. Aquinas, Thomas. In Aristotelis libros de sensu et sensato, De memoria et reminiscentia commentarium. Edited by R.M. Spiazzi. Turin: Marietti, 1949. Pp. viii + 129.

1053. Aquinas, Thomas. The De anima in the Version of William of Moerbeke and the Commentary of St. Thomas Aquinas. Translated by Kenelm Foster and Silvester Humphries with an introduction by Ivo Thomas. New Haven: Yale University Press, 1951. Pp. 504.

1054. Aquinas, Thomas. Quaestiones de anima. Edition of the Latin text by James H. Robb. Toronto: Pontifical Institute of Mediaeval Studies, 1968. Pp. 282.

1055. Averroes (Ibn Rushd). Compendia librorum Aristotelis qui Parva naturalia vocantur. Edited by A.L. Shields and H. Blumberg. Cambridge, Mass.: Mediaeval Academy of America, 1949. Pp. xxxiv + 276.

1056. Averroes (Ibn Rushd). Commentarium magnum in Aristotelis de anima libros. Edited by F. Stuart Crawford. Cambridge, Mass.: Mediaeval Academy of America, 1953. Pp. xxiv + 592.

1057. Averroes (Ibn Rushd). Epitome of Parva naturalia. Translated by Harry Blumberg. Cambridge, Mass.: Mediaeval Academy of America, 1961. Pp. xxii + 180.

Translated from the original Arabic and the Hebrew and Latin versions.

1058. Avicenna (Ibn Sina). Liber de anima seu sextus de naturalibus. 2 vols. Edited by Simone van Riet. Louvain: Editions orientalistes, 1968 (IV & V), 1972 (I-III).

An edition of the medieval Latin translation with an introduction on the psychological doctrines of Avicenna.

1059. Bauer, Hans. Die Psychologie Alhazens. Beiträge zur Geschichte der Philosophie des Mittelalters, 10, 5. Münster: Aschendorff, 1911. Pp. viii + 72.

Summarizes information dealing with the anatomy of the eye and the functions of its parts, and the physiology and psychology of visual perception as found in Alhazen's optical treatise.

1060. Blasius of Parma. Le Quaestiones de anima di Biagio Pelacani da Parma. Edited by Graziella Federici Vescovini. Florence: Olschki, 1974. Pp. 176.

1061. Burley, Walter. De sensibus. Edited by Herman Shapiro and Frederick Scott. Mitteilungen des Grabmann-Institut der Universität München, 13. Munich: Heuber, 1966. Pp. 9.

1062. Giele, Maurice, Fernand van Steenberghen and Bernard Bazán. Trois commentaires anonymes sur le Traité de l'âme d'Aristote. Louvain: Publications universitaires, 1971. Pp. 527.

Discusses an Averroist commentary on Books I and II of the De anima in Oxford, Merton College 275, a semi-Averroist commentary in the same MS and in Clm, MS 9559, and an anti-Averroist commentary on De anima in Bibliothèque Nationale, Paris, Lat. MS 16, 170.

1063. Jackson, Stanley W. "Acedia the Sin and its Relationship to Sorrow and Melancholia in Medieval Times." Bulletin of the History of Medicine, 55 (1981), 172-185.

Traces the fortunes of the concept of acedia through the medieval period both as a vice (spiritual sloth) and as a mental state identified with melancholia.

1063a. John of Jandun. Super libros Aristotelis de anima. Venice, 1587. Reprint. Frankfort: Minerva, 1966.

Bound in the reprint edition with item 614a.

1064. Kennedy, Leonard A. "The De anima of John Sharpe." Franciscan Studies, 29 (1969), 249-270.

Analyzes the content of Sharpe's work, suggesting that it may be the only De anima commentary stemming from England from 1384 to the end of the Middle Ages. Sharpe, a fellow at Queen's College Oxford, 1391-1403, wrote several scientific treatises, including a Physics commentary.

1065. Kitchel, M. Jean "The De potentiis animae of Walter Burley." Mediaeval Studies, 33 (1971), 85-113.

Gives the text of the shortest of three psychological treatises by Burley.

1066. Kurdziałek, Marian. "Gilbert Anglicus und die psychologischen Erörterungen in seinem Compendium medicinae." Sudhoffs Archiv für Geschichte der Medizin und der Naturwissenschaften, 47 (1963), 106-126.

Investigates the sources used by Gilbert in this work, especially in his psychological digression on the soul and its functions and the senses. Concludes Gilbert attempted to draw divergent ideas into a coherent unity.

1067. Maier, Anneliese. "Das Problem der Species sensibiles in medio und die neue Naturphilosophie des 14. Jahrhunderts." Ausgehendes Mittelalter: Gesammelte Aufsätze zur Geistesgeschichte des 14. Jahrhunderts, II. Rome: Edizioni di storia e letteratura, 1967, pp. 419-451.

Points out that although the doctrine of sensible species was accepted with various nuances by thirteenth-century scholastics, it was rejected by many in the fourteenth-century, for example, by Ockham. Nevertheless, the species concept, surprisingly, found approval by two major representatives of the new natural philosophy of the fourteenth century, Buridan and Oresme. Reprinted from Freiburger Zeitschrift für Philosophie und Theologie, 10 (1963), 3-32.

1068. Muckle, J.T. "The Treatise De anima of Dominicus Gundissalinus." Mediaeval Studies, 2 (1940), 23-103.

Gives the text of the treatise. A preface by Gilson points out that Dominicus adapted Avicenna's thought to that of Augustine.

1069. Pack, Roger A. "An Ars memorativa from the Late Middle Ages." Archives d'histoire doctrinale et littéraire du moyen âge, 46 (1979), 221-275.

An edition of a work on the art of memory compiled at Bologna in 1425. Treatises of this kind dealt with the "artificial" memory created by a mental association of material to be memorized with "places," an association in the mind linked to the "places" by "images."

1070. Paniagua, Juan. "La psicoterapía en las obras médicas de Arnau de Vilanova." Archivos Iberoamericanos de historia de la medicina y antropología médica, 15 (1963), 3-15.

Traces the psychotherapeutic themes in Arnald's medical works. Argues that Arnald, as a Galenist, believed in a mutual action between the psyche and the body, and that the physician must consider those "accidents of the soul" which influence the body toward sickness or health.

1071. Philoponus, John. Commentaire sur le De anima d'Aristote par Jean Philopon. Traduction de Guillaume de Moerbeke. Edited with an introduction on the psychology of Philoponus by G. Verbeke. Louvain: Publications universitaires, 1966. Pp. cxx + 172.

1072. Radulph Brito. Der Kommentar des Radulphus Brito zu Buch III De anima. Edited with philosophical and historical introduction by Winfried Fauser.

Beiträge zur Geschichte der Philosophie des Mittelalters, n. f., 12. Münster: Aschendorff, 1974. Pp. viii + 331.

1073. Schneider, Arthur. Die Psychologie Alberts des Grossen. Beiträge zur Geschichte der Philosophie des Mittelalters, 4, 5-6. Münster: Aschendorff, 1903, 1906. Pp. xii + 558.

Argues from an exhaustive treatment of the theological and philosophical features of Albert's explanations of the nature of the soul, its relation to the body and the processes of perception and cognition that Albert attempted to synthesize the Aristotelian doctrines of the De anima and the Parva naturalia with his Augustinian-Neoplatonic sources.

1074. Siger of Brabant. Quaestiones in tertium de anima. De anima intellectiva. De aeternitate mundi. Edited by Bernard Bazán. Louvain: Publications universitaires, 1972. Pp. 231.

1075. Steneck, Nicholas H. "Albert the Great on the Classification and Localization of the Internal Senses." Isis, 65 (1974), 193-211.

Argues that contrary to what has been thought Albert's theory of the internal senses was a unified and unique whole. His classification of these senses was based on the principle of abstraction while their localization depended on the principle of animal spirit. Both principles served to draw seemingly disparate materials into a unity.

1076. Steneck, Nicolas H. "Albert on the Psychology of Sense Perception." Albertus Magnus and the Sciences: Commemorative Essays. Edited by James A. Weisheipl. Toronto: Pontifical Institute of Mediaeval Studies, 1980, pp. 263-290.

Argues that Albert's comprehensive treatment of the senses, basically Aristotelian, as seen in his Summa de creaturis and his commentaries on the De anima, Parva naturalia and De animalibus, made him a pivotal and influential authority.

1077. Tachau, Katherine H. "The Problem of the *species in medio* at Oxford in the Generation after Ockham." *Mediaeval Studies*, 44 (1982), 394-443.

Surveys the negative reactions of several English scholastics as expressed in their *Sentences* commentaries to Ockham's denial of sensible species as central to intuitive cognition. Argues that these scholastics, their views strongly influenced by Bacon and Grosseteste, particularly the latter, generally retained the species notion, finding Ockham's explanation not sufficient to "save the phenomena."

1078. Taddeo of Parma. *Le Questiones de anima di Taddeo da Parma*. Edited by Sofia Vanni Rovighi. Milan: Vita e pensiero, 1951. Pp. xl + 174.

1079. Themistius. *Commentaire sur le traité De l'âme d'Aristote*. *Traduction de Guillaume de Moerbeke*. Edited by G. Verbeke with a study of the utilization of the commentary by S. Thomas Aquinas. Paris: Béatrice-Nauwelaerts, 1957. Pp. xcvii + 322.

An edition of the translation by Moerbeke made in 1267 introduced by essays on Aquinas' use of his commentary and the relation between it and his *De unitate intellectus*.

1080. da Valsanzibio, P. Silvestro. *Vita e dottrina di Gaetano di Thiene*. 2d ed. Padua: Studio filosofico dei FF. MM. Cappuccini, 1949. Pp. xiii + 111.

Sketches the life and career of Gaetano Thienis and lists his works. Associated as student and master with the University of Padua, Gaetano wrote commentaries on logic and Aristotelian texts. Focuses particularly on Gaetano's doctrines in his *De anima* commentary, concluding that his philosophical position was eclectic, neither terminist nor Averroist. He was a moderate Aristotelian with no dogmatic adherence to any school.

## NATURAL HISTORY

### BOTANY

1081. Albertus Magnus. De vegetabilibus libri VII.... Edited by Ernest Meyer and Carl Jessen. Berlin: G. Reimer, 1867. Pp. lii + 752.

1082. Balss, Heinrich. Albertus Magnus als Biologe. Stuttgart: Wissenschaftliche Verlagsgesellschaft, 1947. Pp. 307.

Analyzes the content of Albert's biological thought found in his De animalibus libri XXVI and the De vegetabilibus. Both these commentaries contain original observations Albert made in his native Germany and on his travels elsewhere. Judges Albert to have been an untiring observer although his theory was based on Aristotle and other authorities.

1083. Dioscorides. The Greek Herbal of Dioscorides. Englished by John Goodyer, A. D. 1655. Edited and first printed A. D. 1933 by R.T. Gunther. New York: Hafner, 1959. Pp. ix + 701.

1084. Fischer, Hermann. Mittelalterliche Pflanzenkunde. Munich: Verlag der Münchner Drucke, 1929. Pp. 326.

Surveys botanical thought from the early encylopedists to the herbals of the fifteenth century, including the printed literature. There are special sections on plant illustrations, the cultivation of plants and their medieval European distribution as well as on pharmacological botany. Provides glossaries of medieval and modern plant names and those names of Roman and German origin.

1085. Flood, Jr., Bruce P. "The Medieval Herbal Tradition of Macer floridus." Pharmacy in History, 18 (1976), 62-66.

Summarizes information concerning the Macer floridus, a catalogue of 77 herbs and their medicinal properties, probably written in France between 1070-1112. Based on Pliny, the second-century Roman author Gargalius Martialis and Dioscorides, but not on Arabic sources, the work survives in many MS versions.

1086. Flood, Jr., Bruce P. "Pliny and the Medieval Macer Medical Text." Journal of the History of Medicine and Allied Sciences, 32 (1977), 395-402.

Argues that the material from Pliny in the botanical and medical poem, Macer floridus, may be direct, although Pliny's influence on medical botany was often received from intermediary sources. The text of the Macer floridus is confused and corrupt, and it is difficult to establish a direct use of Pliny.

1087. Frisk, Gösta, ed. A Middle English Translation of Macer Floridus de Viribus Herbarum. Cambridge, Mass.: Harvard University Press, 1949.

1088. Howald, Ernst and Henry Sigerist, eds. Antonii Musae De herba vettonica liber, Pseudo-Apulei herbarius .... Leipzig: Teubner, 1927. Pp. xxiv + 349.

1089. Lambert of St. Omer. Liber floridus. Edited by Alberto Derolez with a preface by I. Strubbe and Alberto Derolez. Ghent: Story-Scientia, 1968. Pp. xlvii + 580 + 114.

1090. Macer floridus De viribus herbarum. Edited by L. Choulant. Leipzig: Voss, 1832. Pp. xii + 220.

1091. Macer floridus De viribus herbarum. Translated with commentary by Roberto Trifogli. Rome: Cossidente, 1958. Pp. 107.

Usually attributed to Odo de Meung (early eleventh century).

1092. Rufinus. The Herbal of Rufinus. Edited by Lynn Thorndike with the assistance of Francis J. Benjamin, Jr. Chicago: University of Chicago Press, 1946. Pp. xliii + 476.

Latin text of the De virtutibus herbarum of c. 1292. Rufinus cites the opinions of authorities such as Dioscorides, Macer floridus and Circa instans.

1093. Schmitt, Charles B. "Theophrastus in the Middle Ages." Viator, 2 (1971), 251-270.

Traces the fortunes of Theophrastus in the Latin West. Although his works, including the two books on botany, were not translated into Latin until 1450, some knowledge of him and his ideas arrived in the West indirectly or in translations attributed to Aristotle. Burley's De vita et moribus philosophorum contains an account of his life and doctrines.

1094. Stannard, Jerry. "A Fifteenth-Century Botanical Glossary." Isis, 55 (1964), 353-367.

Discusses the botanical section of a Huntington Library MS containing a glossary of plant, animal and mineral materials of value to medical practitioners and its sources.

1095. Stannard, Jerry. "Medieval Reception of Classical Plant Names." XIIe congrès international d'histoire des sciences. Revue de synthèse, 89 (1968), 153-162.

Argues that while medieval botanical authors were dependent on classical sources, classical plant names underwent many changes due to scribal carelessness and confusion as to the identity of plants familiar in antiquity, sources of error compounded by the literary genre often used by medieval writers to list and describe plants.

1096. Stannard, Jerry. "Medieval Herbals and their Development." Clio medica, 9 (1974), 23-33.

Characterizes medieval herbals as exhibiting similar formats although with much variation. Before the mid-fifteenth century, the herbal described plants for medical use only; the later fifteenth century saw the development of the _flora_, in which the medical connection was abandoned and all the plants in a locality were described.

1097. Stannard, Jerry. "Botanical Data in Medieval Medical Recipies." _Studies in the History of Medicine_, 1 (1977), 80-87.

Argues that medieval recipe collections contain valuable botanical information. These data are classified in three general groups: descriptions of plants, their nomenclature and their alimentary and other economic uses.

1098. Stannard, Jerry. "Albertus Magnus and Medieval Herbalism." _Albertus Magnus and the Sciences: Commemorative Essays_. Edited by James A. Weisheipl. Toronto: Pontifical Institute of Mediaeval Studies, 1980, pp. 355-377.

Discusses the list of plants in Book VI of Albert's _De vegetabilibus_, analyzes his descriptions and the connection of the plants with medicine and magic.

1099. Strabo, Walahfrid. _Hortulus_. Translated by Raef Payne with commentary by Wilfrid Blunt. Pittsburgh: Hunt Botanical Library, 1966. Pp. xi + 91.

1100. Wickersheimer, Ernest. "Nouveaux texts médiévaux sur le temps de cueillette des simples." _Archives internationales d'histoire des sciences_, 29 (1950), 342-355.

Reports two texts, one in German, which give the time of year, month by month, for the gathering of herbs. Both are dependent on an _Epistola Ypocratis ad Alexandrum de tempore herbarum_ written after the twelfth century.

1101. Wimmer, Josef. _Deutsches Pflanzenleben nach Albertus Magnus (1197-1280)_. Halle a. S.: Verlag der Buchhandlung des Waisenhauses, 1908. Pp. 77.

Discusses the wild and cultivated plants mentioned by Albert in his commentary on the pseudo-Aristotelian _De vegetabilibus_. Albert's extensive travels through German lands enabled him to report first-hand observations of German plants in the thirteenth century.

# ZOOLOGY

1102. Albertus Magnus. De animalibus libri XXVI. Edited by Hermann Stadler. Beiträge zur Geschichte der Philosophie des Mittelalters, 15 and 16. Münster: Aschendorff, 1916, 1920. Pp. xxvi + 892, xxi + 772.

Vol. 15 contains an edition of Books I-XII; Vol. 16, Books XIII-XXVI.

1103. Aristotle. De generatione animalium. Translatio Guillelmi de Moerbeka. Edited by H.J. Drossaart Lulofs. Aristoteles latinus, XVII, 2. Bruges: Desclée de Brouwer, 1966. Pp. xxxii + 270.

* Balss, Heinrich. Albertus Magnus als Biologe. Stuttgart: Wissenschaftliche Verlagsgesellschaft, 1947. Pp. 307.

Cited as item 1082.

1104. Balss, Heinrich. "Die Tausendfüssler, Insekten und Spinnen bei Albertus Magnus." Sudhoffs Archiv für Medizin und der Naturwissenschaften, 38 (1954), 303-322.

Discusses the millipedes, insects and spiders described in the last book of Albert's De animalibus, arranged by classes and orders.

1105. Carmody, Francis J., ed. Physiologus latinus. Éditions préliminaires, versio B. Paris: E. Droz, 1939.

1106. Carmody, Francis J., ed. "Physiologus latinus, versio Y." University of California Publications in Classical Philology, 12 (1941), 95-134.

1107. Curley, Michael J. Physiologus. Austin: University of Texas Press, 1979. Pp. xliii + 92.

An English translation based on two versions of the Latin Physiologus edited by Francis Carmody. See items 1105, 1106.

1108. Frederick II. The Art of Falconry, being the De arte venandi cum avibus of Frederick II of Hohenstaufen.

Translated and edited by Casey A. Wood and F. Marjorie Fyfe. Palo Alto, 1943. Reprint. Boston: Charles T. Branford Co., 1955. Pp. cvi + 637.

1109. Gerhardt, Mia I. "Knowledge in Decline: Ancient and Medieval Information on 'Inkfishes' and their Habits." Vivarium, 4 (1966), 144-175.

Traces the fortunes of information concerning cephalopods from antiquity to Albertus Magnus. Familiar to the ancient Mediterranean world, cephalopods were described by Aristotle, but knowledge about them suffered at the hands of Pliny and Isidore. The return of Aristotle's zoological works to the West and Albert's first-hand descriptions revived more accurate information concerning the creatures.

1109a. Gerhardt, Mia I. "Zoologie médiévale: Préoccupations et procédés." Methoden in Wissenschaft und Kunst des Mittelalters. Edited by Albert Zimmermann and prepared for the press by Rudolf Hoffmann. Berlin: de Gruyter, 1970, pp. 231-248.

Describes varied types of literature revealing different aspects of the medieval concern with animals --the religious and allegorical (the Physiologus and the Bible), a utilitarian focus (medical use of animals and animal products, treatises on hunting and falconry), the literary and learned tradition introduced by Aristotle's zoological works and employed by the thirteenth-century encyclopedists and the tradition based on actual experience and observation (Frederick II and Albertus Magnus).

1110. Killermann, S. Die Vogelkunde des Albertus Magnus (1207-1280). Regensburg: G.J. Manz, 1910. Pp. vi + 100.

Abstracts information concerning the anatomy, reproduction and behavior of birds from Albert's De animalibus. Albert mentioned over a hundred different birds taken from Aristotle and Pliny, but also from his own observations. Many birds, particularly those native to Germany, were listed by Albert for the first time.

1111. McCollough, Florence. Medieval Latin and French Bestiaries. Rev. ed. Chapel Hill: University of North Carolina Press, 1962. Pp. 212.

Traces the development of the Latin bestiary from its origin in the Greek Physiologus, and discusses the content of four major French bestiaries based on the Latin tradition--those by Phillipe de Thaon, Gervase, Guillaume le Clerc, and two versions probably by Pierre de Beauvaise.

1112. Oggins, Robin S. "Albertus Magnus on Falcons and Hawks." Albertus Magnus and the Sciences: Commemorative Essays. Edited by James A. Weisheipl. Toronto: Pontifical Institute of Mediaeval Studies, 1980, pp. 441-662.

Points out that half of the section of Albert's De animalibus describing birds is devoted to hawks and falcons. These birds are treated in great detail, combining written sources, information from falconers and Albert's own observations; it is possible to identify many of the birds and to confirm Albert's account of their behavior.

1113. Scott, Frederick and Herman Shapiro. "Jean Buridan's De motibus animalium." Isis, 58 (1967), 533-552.

An edition prepared from two MS versions of this commentary. The work is an excellent clarification of Aristotle's views on the movement of animals in which Buridan presents his own opinions in opposition to Aristotle.

1114. Scott, Frederick and Herman Shapiro. "Walter Burley's Commentary on Aristotle's De motu animalium." Traditio, 25 (1969), 171-190.

An edition of the text.

1115. Théodoridès, Jean. "Orient et occident au moyen âge: L'oeuvre zoologique de Frederic II Hohenstaufen." Oriente e occidente nel medioevo: Filosofia e scienze. Rome: Accademia nazionale dei Lincei, 1971, pp. 549-567.

Discusses the zoological interests of Frederick II, focusing on his De arte venandi cum avibus, written between 1244 and 1250. In this work, Frederick treats all aspects of falcons from anatomy, diseases, mechanics of flight to the capture, care and training of young birds. Frederick relies on facts he has observed, and his attitude toward authority, especially Aristotle, is frequently critical. The illustrations in the text were likely done by an artist under Frederick's direction.

1116. White, T.H., ed. The Book of Beasts, being a Translation from a Latin Bestiary of the Twelfth Century. New York, 1955. Reprint. London: Jonathan Cape, 1956. Pp. 296.

## GEOGRAPHY AND CARTOGRAPHY

1117. d'Ailly, Pierre. Ymago mundi. 3 vols. Edited with a French translation by Edmond Buron. Paris: Maisonneuve Frères, 1930. Pp. 823.

Texts of four cosmological and geographical treatises.

1118. Beazley, Charles R. The Dawn of Modern Geography: A History of Exploration and Geographical Science. 3 vols. London, 1897-1903. Reprint. New York: Peter Smith, 1949.

Provides an account of geographical travels--exploration and conquest, commercial and diplomatic, those undertaken by pilgrims and missionaries, as well as a survey of the geographical accounts of Latin, Arabic and Hebrew writers and map-makers.

1119. von den Brincken, Anna-Donothee. "Mappa mundi und Chronographia: Studien zur imago mundi des abendländischen Mittelalters." Deutsches Archiv für Erforschung des Mittelalters, 24 (1968), 118-186.

Characterizes the medieval world map as an encyclopedic and historical scheme of the world based on ancient and biblical sources. Providing an overview of the chief places on the earth, it was not intended for practical geographical purposes. It was rather a view of world history embedded in the Passion.

1119a. von den Brincken, Anna-Dorothee. "... ut describeretur universus orbis: Zur Universalkartographie des Mittelalters." Methoden in Wissenschaft und Kunst des Mittelalters. Edited by Albert Zimmermann and prepared for the press by Rudolf Hoffmann. Berlin: de Gruyter, 1970, pp. 249-278.

Argues that the above command of the Emperor Augustus (in Luke) can serve as a description of the medieval world map, an imago mundi, a depiction of the major events and localities in a human history related to God. Treats types of world maps--the dominant T-map with its three-fold division of the world (derived from antiquity) and variations of it. Scientific cartography, which appeared at the end of the M. A., was a contribution of the Arabs.

1120. Destombes, Marcel, ed. Monumenta cartographica vetustioris aevi, A. D. 1200-1500, I: Mappemondes. Amsterdam: Israel, 1964. Pp. xxxii + 322.

1121. Flint, Valerie I.J. "Honorius Augustodunensis Imago mundi." Archives d'histoire doctrinale et littéraire du moyen âge, 49 (1982), 7-153.

Provides an edition of Honorius of Autun's cosmological and geographical work.

1122. Kimble, George H.R. Geography in the Middle Ages. London, 1938. Reprint. New York: Russell & Russell, 1968. Pp. vi + 272.

A survey of geographical knowledge from late antiquity to the end of the Middle Ages.

1123. Klauck, K. "Albertus Magnus und die Erdkunde." Studia Albertina. Beiträge zur Geschichte der Philosophie des Mittelalters, supp. 4 (1952), 234-248.

Summarizes Albert's geographical ideas as seen in his De natura loci. The work is divided into three tracts: the first two are devoted to general geography and treat the earth and its surface as a whole, its shape and size and position in the universe. The remainder of the treatise is concerned with physical geography, the habitable portions of the earth, climatology and morphological questions.

1124. Moody, Ernest A. "John Buridan on the Habitability of the Earth." Studies in Medieval Philosophy, Science, and Logic: Collected Papers, 1933-1969. Berkeley: University of California Press, 1975, pp. 111-126.

Points out that the arguments Buridan adduces in his De caelo commentary as to the habitability of the entire earth (which he rejects) are based on physical and mechanical explanations.

1125. Stevens, Wesley M. "The Figure of the Earth in Isidore's De rerum natura." Isis, 71 (1980), 268-277.

Argues that the diagram of the earth in Isidore's work (a rota terrarum) does represent the world as a sphere and not a disk. His description of the five climates on the earth's surface has led to a similar confusion with the main circles of the celestial sphere. Points out that many so-called errors in medieval drawings of the earth are really due to differences in perspective.

1126. Tilmann, Jean Paul. An Appraisal of the Geographical Works of Albertus Magnus and his Contributions to Geographical Thought. Michigan Geographical Publication, 4. Ann Arbor: Department of Geography, University of Michigan, 1971. Pp. xi + 190.

Analyzes Albert's De natura locorum and provides an English translation of the text. Argues that Albert's treatise, written between 1248-1252, summarizes the geographical knowledge of his time, but relies neither on the usual geographical authorities of the early M. A. nor the Bible, but rather newly-translated materials and Arabic interpretations.

1127. Wright, John K. "Notes on the Knowledge of Latitudes and Longitudes in the Middle Ages." Isis, 5 (1923), 75-98.

Argues that information concerning terrestrial longitudes and latitudes in the Latin West was derived from Arabic and Hellenistic sources, but that interest in these coordinates was virtually confined to astronomical-astrological purposes. Although the astronomical positions of many places were known, as was Ptolemy's use of parallels of latitude and places on each parallel, this knowledge was put to no practical geographical use.

1128. Wright, John K. The Geographical Lore of the Time of the Crusades: A Study in the History of Medieval Science and Tradition in Western Europe. New York, 1925. Reprint. New York: Dover Publications, 1965. Pp. xxi + 563.

Describes the geographical, cosmological and meteorological ideas prevalent in the Latin West before 1100 and from 1100-1250. These notions were dependent on inherited traditions--the Bible, the Church Fathers and classical antiquity--but also included information from observations made by eyewitnesses.

## MINERALOGY

1129. Albertus Magnus. Book of Minerals. Translated by Dorothy Wyckoff. Oxford: Clarendon Press, 1967. Pp. xlii + 309.

1130. Avicenna (Ibn Sina). De congelatione et conglutinatione lapidum. Edited and translated into English by E.J. Holmyard and D.C. Mandeville. Paris: Paul Geuthner, 1927.

The Latin and Arabic texts of what is partly a direct translation and partly a resume of sections of the al-Shifā. The work is divided into three parts: De congelatione et conglutinatione lapidum, De causa montium and De quatuor speciebus corporum mineralium.

1131. Evans, Joan. Magical Jewels of the Middle Ages and the Renaissance, particularly in England. Oxford, 1922. Reprint. New York: Dover Publications, 1976. Pp. 264 + [4].

Traces lapidary traditions in Europe and discusses types of lapidaries. Provides the text of several works in this genre, including that of Evax. See item 1133.

1132. Evans, Joan and Mary Serjeantson. English Mediaeval Lapidaries. London, 1933. Reprint. London: Oxford University Press, 1960. Pp. xii + 205.

Texts of several English lapidaries.

1133. Halleux, Robert. "Damigéron, Evax et Marbode: L'heritage alexandrin dans les lapidaires médiévaux." Studi medievali, 15 (1974), 327-347.

Traces the origins and sources of two versions of the lapidary of Evax. Argues that Marbode versified the alphabetical version. The alphabetical Evax has interpolations from the alphabetical Latin Dioscorides, itself possibly from a Salernitan ambiance.

1134. Kitson, Peter. "Lapidary Traditions in Anglo-Saxon England, Part I: The Background, the Old English Lapidary." Anglo-Saxon England, 7 (1978), 9-60.

Examines the background of European and English lapidary traditions arising from classical sources, from magic and medical lore and from Biblical exegesis. Provides the text of an eleventh-century Anglo-Saxon lapidary, the oldest vernacular lapidary in western Europe.

1135. Klein-Franke, Felix. "The Knowledge of Aristotle's Lapidary during the Latin Middle Ages." Ambix, 17 (1970), 137-142.

Treats the problem of the transmission of this pseudo-Aristotelian work to Europe and the date of its arrival in the West. One can say with certainty that it was used by Marbode of Rennes (d. 1123).

1136. Riddle, John M. and James A. Mulholland. "Albert on Stones and Minerals." Albertus Magnus and the Sciences: Commemorative Essays. Edited by James A. Weisheipl. Toronto: Pontifical Institute of Mediaeval Studies, 1980, 203-234.

Maintains that Albert was a pioneer in the science of mineralogy, a field in which he had no precursors. His work on minerals not only supplied a theoretical base for chemical technology, but also drew on the practical knowledge of lapidarists, alchemists, miners and other artisans and professionals.

1137. Studer, Paul and Joan Evans. Anglo-Norman Lapidaries. Paris: Champion, 1924. Pp. xx + 404.

Texts of several lapidaries. These all rely on the usual medieval lapidary tradition based on Pliny, Isidore, Alexandrian lapidaries and Marbode of Rennes.

1138. Wyckoff, Dorothy. "Albertus Magnus on Ore Deposits." Isis, 49 (1958), 109-122.

Characterizes Albert's Mineralium libri V as the most original of his scientific works, being neither a paraphrase nor a commentary. Deals with passages on ore genesis in which Albert provides explanations of the formation of ores based on chemical processes.

# TECHNOLOGY

## GENERAL

1139. Gille, Bertrand. "Le moyen âge en occident (Ve siècle-1350)." Histoire générale des techniques, I. Edited by Maurice Daumas. Paris: Presses universitaires de France, 1962, pp. 427-597.

Covers a wide range of technical methods and devices used in transportation, agricultural exploitation, warfare, mining, the preparation of materials, techniques of manufacture or construction. Treats problems of origins and transmission, social organization and the relationship of technology and science.

1140. Gille, Bertrand. "Technological Developments in Europe: 1100-1400." The Evolution of Science. Edited by Guy S. Métraux and François Crouzet. New York: New American Library, 1963, pp. 168-219.

Surveys medieval technology from three viewpoints: the technology itself (power sources, machines, implements in various fields), the problems associated with the origin and transmission of technological ideas, and the social, economic and legal consequences of the medieval mechanical civilization. Translated from Cahiers d'histoire mondiale, 3.

1141. Gimpel, Jean. The Medieval Machine: The Industrial Revolution of the Middle Ages. New York: Holt, Rinehart & Winston, 1976. Pp. ix + 274.

Surveys the technological developments of the period. The author maintains that the inventiveness and creativity of the M. A. represented a genuine industrial revolution.

1142. Feldhaus, Franz M. Die Technik der Antike und des Mittelalters. Potsdam, 1930. Reprint. Hildesheim: Olms, 1970. Pp. 442.

Medieval technology in Part IV (pp. 219-425). "Erfinder und Ingenieure des Mittelalters." A survey of miscellaneous techniques and devices.

1143. Ploss, Emil Ernst. Ein Buch von alter Farben: Technologie der Textilfarben im Mittelalter mit einer Ausblick auf die festen Farben. Heidelberg: Heinz Moos, 1962. Pp. 168.

Discusses the history of dyes, dyeing and textile printing in the Middle Ages.

1144. Salzmann, L.F. English Industries of the Middle Ages. London: Constable, 1913. Pp. 260.

Contains chapters on mining, quarrying, metalworking, pottery and tile work, textiles, leatherworking and brewing.

1145. Singer, Charles, E.J. Holmyard, A.R. Hall and Trevor I. Williams. A History of Technology, II: The Mediterranean Civilizations and the Middle Ages, c. 700 B. C. to c. A. D. 1500. London: Oxford University Press, 1956. Pp. lix + 802.

Contains essays on many aspects of the technological culture of the medieval period, including areas such as primary production, manufacture, material civilization, transport, practical mechanics and chemistry. Individual essays are followed by a bibliography.

1146. White, Jr., Lynn. Medieval Technology and Social Change. Oxford, 1962. Paperback reprint. Oxford: Oxford University Press, 1976. Pp. x + 194.

Examines three areas of technological innovation and their impact on society: the stirrup's advent; the introduction of the heavy plow, the horse collar and the three-field system of crop rotation; the exploitation of power sources and the development of machine design.

1147. White, Jr., Lynn. Medieval Religion and Technology: Collected Essays. Berkeley: University of California Press, 1978. Pp. xxiv + 360.

A collection of previously published articles. Includes items 1194, 1195, 1198, 1200, 1201, 1202, 1203 and 1204.

## TECHNOLOGY: SPECIFIC TOPICS

1148. Alessio, Franco. "La filosofia e le artes mechanicae nel secolo XII." Studi medievali, 6 (1965), 71-155.

Discusses the philosophical suppositions which determined attitudes toward the mechanic arts in the twelfth century, particularly in the works of the two great classifiers of the sciences. Hugh of St. Victor and Dominicus Gundissalvus. To Hugh, the mechanic arts were a means of access to God, but in Dominicus, these artes were bound up with the human social order.

1149. Allard, Guy H. "Les arts mécaniques aux yeux de l'idéologie médiévale." Cahiers d'études médiévales, VII. Les arts mécaniques au moyen âge. Edited by Guy H. Allard and Serge Lusignan. Paris: J. Vrin, 1982, pp. 13-31.

Argues that the attitude of the intellectual elite toward the mechanic arts was one of denigration and fear. These arts served as a source of metaphor which both distanced and concealed their nature. They were a threat to a hierarchically-ordered society, first due to their connection with pagan, occult practices and, later, due to their natural association with the increasing economic power of the commercial class.

1150. Balmer, R.T. "The Operation of Sand Clocks and Their Medieval Development." Technology and Culture, 19 (1978), 615-632.

Maintains that the sand clock appears only in the late medieval period; the earliest known depiction is a Lorenzetti fresco in Siena of 1338-1339. Given their similarity to the ancient water clock, their late appearance is remarkable. They were commonly used on shipboard for navigational purposes and to time seamen's watches. Medieval and early modern sand clocks could not be graduated.

1151. Beaujouan, Guy. L'interdépendence entre la science scholastique et les techniques utilitaires (XIIe, XIIIe et XIVe siècles). Paris. Palais de la Découverte, 1957. Pp. 20.

Argues that there was a fruitful interaction between theoretical science and the work of artisans and technicians, especially after the mid-thirteenth century, as attested by numerous examples.

1152. Beaujouan, Guy. "Reflexions sur les rapports entre théorie et pratique au moyen âge." The Cultural Context of Medieval Learning. Edited by John E. Murdoch and Edith D. Sylla. Dordrecht: Reidel, 1975, pp. 437-484.

Points out the circumspection necessary in the delicate problem of the interrelation between theory and practice in the Middle Ages, especially the theoretical expertise of those in the practical arts. Illustrated primarily from the fields of architecture and navigation.

1153. Birkenmajer, Aleksander. "Zur Lebensgeschichte und wissenschaftlichen Tätigkeit von Giovanni Fontana (1395?-1455?)." Études d'histoire des sciences et la philosophie du moyen âge. Edited by Aleksandra Maria Birkenmajer and Jerzy B. Korolec. Studia Copernicana, 1. Warsaw: Polish Academy of Sciences, 1970, pp. 529-549.

Traces the life and career of Fontana. Originally from Venice, he studied at Padua where he was a doctor of medicine. Fontana wrote works on the practical applications of mathematics and physics. Reprinted from Isis, 17 (1932), 34-53.

1154. Boyer, Marjorie Nice. "Mediaeval Suspended Carriages." Speculum, 34 (1959), 359-366.

Supplies evidence that the carriage with suspension, assumed to have originated with the sixteenth-century coach, was known in France in the late Middle Ages.

1155. Boyer, Marjorie Nice. "Medieval Pivoted Axles." *Technology and Culture*, 1 (1960), 128-138.

Adduces evidence to show that four-wheeled vehicles with pivoted axles, known to the Romans, were also used in the Middle Ages, despite opinion to the contrary. See item 1167.

1156. Burnham, John M., ed. *A Classical Technology*. Boston: Gorham Press, 1920. Pp. 170.

An edition of the late eighth- or early ninth-century treatise on the technology of materials called the *Compositiones ad tingenda musiva* or *Compositiones variae*.

1157. *Compositiones ad tingenda musiva* [Codex Lucensis 490]. Edited and translated by Hjalmar Hedfors. Uppsala: Almquist & Wiksells, 1932. Pp. xviii + 226.

1158. Fitchen, John. *The Construction of Gothic Cathedrals: A Study of Medieval Vault Erection*. Oxford: Clarendon Press, 1961. Pp. xxi + 344.

Combines the separate provinces of architect and engineer in an account of the procedures, methods and equipment which might have been employed in the erection of Gothic cathedrals, especially in France.

1159. Forbes, R.J. "Metallurgy and Technology in the Middle Ages." *Centaurus*, 3 (1953-54), 49-57.

Points out that classical traditions continued in medieval metallurgy, but considerable progress was made, especially in the use of water power in mining procedures, and above all, in the development of cast-iron. Techniques of mass production and standardization were applied to cast-iron products, including those for military purposes.

1160. Ganzenmüller, Wilhelm. "Ein unbekanntes Bruchstück der *Mappae clavicula* aus dem Anfang des 9. Jahrhunderts." *Mitteilungen zur Geschichte der Medizin, der Naturwissenschaft und der Technik*, 40 (1941), 1-15.

Announces the discovery of a two-page fragment of the *Mappae clavicula* from Klosterneuburg from the last third of the ninth century and earlier than the oldest known MS. See item 1184.

1161. George, Nadine F. "Albertus Magnus and Chemical Technology in a Time of Transition." Albertus Magnus and the Sciences: Commemorative Essays. Edited by James A. Weisheipl. Toronto: Pontifical Institute of Mediaeval Studies, 1980, pp. 235-261.

Argues, chiefly from Albert's commentary on the Meteorology and his own work on minerals, that Albert played an important role in transmitting information on chemical processes, often based on his own observations.

1162. Gille, Bertrand. Esprit et civilisation techniques au moyen âge. Paris: Palais de la Découverte, 1952. Pp. 25.

Traces the development of a medieval technological spirit as evidenced by radical innovations in the high Middle Ages in agricultural methods, ship design, mining techniques, instruments of war and mechanical time pieces. The most striking feature of the mechanical revolution of this period was the wide development of mechanisms using primarily hydraulic power. Discusses also the interaction between social, economic, political and human factors and the medieval technological attitude.

1163. Gille, Bertrand. "La naissance du système bielle-manivelle." Techniques et civilisations, 2 (1952), 42-46.

Points out the difficulties in tracing the origins of the crank with connecting rod; there are no texts and few illustrations. The earliest depiction of a connecting rod dates from the mid-ninth century, while illustrations of the crank and connecting rod date from between 1421-1434. There are no other examples for the remainder of the fifteenth century.

1164. Gille, Bertrand. "Le moulin à eau: Une révolution technique médiévale." Techniques et civilisations, 3 (1954), 1-15.

Traces the spread of the water mill throughout Europe from its origin in the Mediterranean area in late antiquity. Discusses types of mills (under-and over-shot), ship-mills, tidal mills and the use of dams and watercourses. The water wheel produced a continuous circular motion, but a chief feature of the medieval technological revolution was the addition of a system of cams to the water wheel which, with the weight of the apparatus and a spring, converted rotary motion to a back-and-forth one.

1165. Gille, Bertrand. "Les problèmes de la technique minière au moyen âge." Revue d'histoire des mines et de la metallurgie, 1 (1969), 279-297.

Discusses the sources of information concerning medieval mining: mining regulations and laws (deriving from a common center in Germany), pictorial representations of miners and mining methods, and evidence offered by surviving excavations. Treats also mining techniques in those mines which had galleries opening off primary shafts, describes the tools employed and the methods adopted to meet problems of illumination, ventilation and the removal of water.

1166. Cillmor, Carroll M. "The Introduction of the Traction Trebuchet into the Latin West." Viator, 12 (1981), 1-8.

Argues that there is evidence that the traction trebuchet may have arrived in the Latin West in the ninth century, either through Moslem Spain or Byzantine Italy, rather than in the mid-twelfth century.

1167. Hall, A. Rupert. "More on Medieval Pivoted Axles." Technology and Culture, 2 (1961), 17-22.

Supplies evidence from Guido da Vigevano's Texaurus which shows that the pivoted axle was known in 1335 (date of the Texaurus). Suggests that the matter of fact treatment in Guido's treatise implies it was a familiar mechanism.

1168. Hall, A. Rupert. "Guido's Texaurus, 1335." Pre-Modern Technology and Science. Edited by Bert S. Hall and Delmo C. West. Malibu: Undena Publications, 1976, pp. 11-52.

Discusses the *Texaurus regis Francie acquisitionis terre sancte de ultra mare ...* of Guido de Vigevano, a text primarily on the military technology which might be employed on crusade.

1169. Hall, Bert S. "Giovanni de' Dondi and Guido da Vigevano: Notes Toward a Typology of Medieval Technological Writings." *Machaut's World*: *Science and Art in the Fourteenth Century*. Edited by Madeleine Pelner Cosman and Bruce Chandler. New York: New York Academy of Sciences, 1978, pp. 127-142.

Distinguishes two types of fourteenth-century technological treatise. That of de' Dondi (an older type) was concerned with the design and construction of a precision instrument, while Guido's treatise, written to advise Philip VI with his plans for a Crusade, contains realistic and imaginative designs of devices for warfare intended for a general public.

1170. Hall, Bert S. "*Der Meister sol auch kennen schreiben und lesen*: Writings about Technology, ca. 1400-1600 A. D. and their Cultural Implications." *Early Technologies*. Edited by Denise Schmant-Besserat. Malibu: Undena Publications, 1979, pp. 46-58.

Points out that medieval technology, previously the province of the illiterate or semi-literate, underwent a quantitative change in the fifteenth century when written treatises began to express complicated ideas. Discusses the spread of technical information due to the invention of printing, the audiences for these works and their interrelationship with art.

1171. Hall, Bert S. "Production et diffusion de certain traités de techniques au moyen âge." *Cahiers d'études médiévales*, VII. *Les arts mécaniques au moyen âge*. Edited by Guy H. Allard and Serge Lusignan. Paris: J. Vrin, 1982, pp. 147-170.

Examines works on technology as to their purpose, audience and connection with ancient technical writings. Points out the marked change in technical works in the fourteenth century where treatises were often written by physicians, were concerned with military machines and were destined for princely courts.

1172. Kreutz, Barbara M. "Mediterranean Contributions to the Medieval Mariner's Compass." *Technology and Culture*, 14 (1973), 367-383.

Argues that while the properties of a floating magnetic needle in a bowl was a Chinese invention, the appearance of the dry pivoted needle with windrose or compass card in Europe by the beginning of the fourteenth century needs to be accounted for. Argues that the 16-point windrose, departing from the classical tradition of the 12 winds, may go back to an ancient Mediterranean tradition of divinatory or ritual use of a floating needle in a bowl with a 16-point division of the horizon.

1173. Kyeser, Conrad (aus Eichstätt). *Bellifortis*. 2 vols. Edited by Götz Quarg. Dusseldorf: V. D. I.-Verlag, 1967.

1174. Muendel, John. "The Horizontal Mills of Medieval Pistoia." *Technology and Culture*, 15 (1974), 194-225.

Argues that the large number of mills for grinding grain in Pistoia, operated by the numerous streams of the region, were primarily of the horizontal type. Describes the structure of these mills. They may have also been dominant in other parts of Tuscany, and may have been in Pistoia by the first century A. D. or earlier.

1175. Prager, Frank D. and Gustina Scaglia. *Mariano Taccola and his Book De ingeneis*. Cambridge, Mass.: M. I. T. Press, 1972. Pp. xii + 230.

An edition of selected drawings with texts and notes from this work by the fourteenth-century Sienese artist and engineer. See item 1189.

1176. Sanfaçon, Roland. "Le rôle des techniques dans les principales mutations de l'architecture gothique." *Cahiers d'études médiévales*, VII. *Les arts mécaniques au moyen âge*. Edited by Guy H. Allard and Serge Lusignan. Paris: J. Vrin, 1982, pp. 93-129.

Treats the innovations introduced in Gothic buildings of the twelfth and thirteenth centuries, the details of the tools used and the allocation of tasks in the work place, as well as those technical features of thirteenth-century work which influenced the later Gothic.

1177. Shelby, Lon R. "Medieval Mason's Tools: The Level and Plumb Rule." Technology and Culture, 2 (1961), 127-130.

Stresses the use of medieval miniatures as a source of information on building tools. Discusses the level (straight edge with plumb bob) used in laying a horizontal course of stone, and the plumb rule, employed to check the straightness of vertical walls.

1178. Shelby, Lon R. "The Role of the Master Mason in Mediaeval English Building." Speculum, 39 (1964), 387-403.

Discusses the manifold activities of the master mason who designed buildings in stone, handled the administrative details, served as a building contractor and supervised the progress of construction.

1179. Shelby, Lon R. "Mediaeval Mason's Tools. II: Compass and Square." Technology and Culture, 6 (1965), 236-248.

Shows that the medieval compass was a simple tool of the divider type; the best example of its use in building practice is seen in Villard de Honnecourt. The square, likewise simple, was essential for setting out construction work. Discusses the various theories that account for tools with curved arms or tapered sides which appear in medieval miniatures.

1180. Shelby, Lon R. "Setting out the Key Stones of Pointed Arches: A Note on Medieval 'Baugeometrie'." Technology and Culture, 10 (1969), 537-548.

Discusses the puzzling pages devoted to stereotomy in the thirteenth-century manuscript sketchbook of Villard de Honnecourt. See item 1191.

1181. Shelby, Lon R. "The Education of Medieval English Master Masons." Mediaeval Studies, 32 (1970), 1-26.

Argues that it is unlikely that the formal education of master masons extended to more than simple literacy. Thus masons had no direct contact with university learning nor would this have given them the required expertise. However, they did have contact with high scholastic culture through constant communication with their patrons. They learned the necessary techniques from relatives, apprenticeship to a master and by long experience working in all aspects of their craft.

1182. Shelby, Lon R. "Mediaeval Mason's Templates." Journal of the Society of Architectural Historians, 30 (1971), 140-154.

Discusses the construction and use of templates in the design and production of mouldings and profiles in carved stone. Since medieval architectural drawings were not detailed, templates played a large role in design. Information concerning templates can be found in financial records and in the Sketchbook of Villard de Honnecourt.

1183. Shelby, Lon R. "The Geometrical Knowledge of Mediaeval Master Masons." Speculum, 47 (1972), 395-421.

Argues that the art of geometry of medieval masons meant the skill to treat design and building problems in terms of a few geometrical figures which could be manipulated through a series of definite steps.

1184. Smith, Cyril Stanley and John G. Hawthorne, trans. Mappae clavicula: A Little Key to the World of Medieval Techniques. Transactions of the American Philosophical Society, n. s., 64, part 4. Philadelphia: American Philosophical Society, 1974. Pp. 128.

An annotated translation with a discussion of the Mappae clavicula as a source for medieval technology.

1185. Sternnagel, Peter. Die artes mechanicae im Mittelalter: Begriffs-und Bedeutungsgeschichte bis zum Ende des 13. Jahrhunderts. Kallmünz: Lassleben, 1966, Pp. 153.

Traces the development of the concept of the mechanical arts with particular attention to linguistic factors, social considerations and attitudes toward hand work. Discusses the treatment of these artes vis-à-vis the liberal arts from Erigena through their position in the classification scheme of Hugh of St. Victor to their place in the thought of Albertus Magnus and Aquinas.

1186. Taylor, E.G.R. "Mathematics and the Navigator in the Thirteenth Century." Journal of the Institute of Navigation, 13 (1960), 1-12.

Argues that the appearance of the maritime chart drawn to scale with its network of rhumb-lines required trigonometric knowledge for the erection of tables and for finding a graphical solution to the nautical triangle.

1187. Theophilus. De diversis artibus. Edited with English translation by C.R. Dodwell. London: Thomas Nelson, 1961. Pp. lxxx + 178.

1188. Theophilus. On Divers Arts: The Treatise of Theophilus. Translated by John G. Hawthorne and Cyril Stanley Smith. Chicago: University of Chicago Press, 1963. Pp. xxxv + 216.

1189. Thorndike, Lynn. "Marianus Jacobus Taccola." Archives internationales d'histoire des sciences, 8 (1955), 7-26.

Discusses previous notices, various MSS and relations with similar works, of the profusely illustrated De machinis libri decem (1449) of Taccola, a treatise on machines and mechanical devices including engines of war.

1190. Timmermann, Gerhard. "Technik zur Zeit der Hanse." Technikgeschichte, 36 (1969), 265-276.

Discusses the technology of the ships of the Hansa and the nature of their cargoes.

1191. Villard de Honnecourt. The Sketchbook of Villard de Honnecourt. 2d rev. ed. Edited by Theodore Bowie. Bloomington, Ind.: Indiana University Press, 1962. Pp. 80.

1192. Vowles, Hugh P. "An Inquiry into Origins of the Windmill." Transactions of the Newcomen Society, 11 (1930-31), 1-13.

Argues, based on Hero's "windmill" in his Pneumatics, that the windmill was Hellenistic in origin. The device was transferred to Persia from whence it came back to the West by the tenth and eleventh centuries.

1193. Wailes, Rex. "Tidemills in England and Wales." Transactions of the Newcomen Society, 19 (1938-39), 1-33.

Surveys sites of water mills using tidal water as a power source. Some date back to the medieval period and some are still in use.

1194. White, Jr., Lynn. "Technology and Invention in the Middle Ages." Medieval Religion and Technology: Collected Essays. Berkeley: University of California Press, pp. 1-22.

Argues that medieval Europe used technological skills from many areas, and that knowledge from Greek and Roman sources were subject to many improvements. The technological enterprise of the medieval world is revealed in revolutionary changes in agriculture, in practical mechanics and in the wide use of water- and windmills which substituted mechanical power for human. This technological advance arose from the ideals of Christian theology which stressed not only the dignity of labor but also the intrinsic worth of the individual and the need to release him from drudgery. Reprinted from Speculum, 15 (1940), 141-159.

1195. White, Jr., Lynn. "Natural Science and Naturalistic Art in the Middle Ages." Medieval Religion and Technology: Collected Essays. Berkeley: University of California Press, 1978, pp. 23-41.

Argues that the awakening of interest in nature in the twelfth century was matched by a change in artistic expression. Both mirror a transformation from the view of nature as "symbolic-subjective" to "naturalistic-objective." Reprinted from American Historical Review, 52 (1947), 421-435.

1196. White, Jr., Lynn. "Eilmer of Malmesbury, An Eleventh Century Aviator: A Case Study of Technological Innovation, its Context and Tradition." Medieval Religion and Technology: Collected Essays. Berkeley: University of California Press, 1978, pp. 59-73.

Discusses the motivation for the experience of Eilmer who, probably between 1000 and 1010, became the first European to fly a short distance by means of wings attached to his hands and feet. While Eilmer may have heard of a similar Arabic attempt in ninth-century Spain, there was a technological atmosphere in Anglo-Saxon England which could have encouraged an original effort. Reprinted from Technology and Culture, 2 (1961), 97-111.

1197. White, Jr., Lynn. "What Accelerated Technological Progress in the Western Middle Ages?" Scientific Change. Edited by A.C. Crombie. New York: Basic Books, 1963, pp. 272-291.

Argues that the enthusiasm of the Latin West for technology as compared with Byzantium and the Moslem East, was based on the following factors: a Celtic heritage of craftsmanship, the openness of Europe to change, the appearance of a new attitude toward the environment due to the advent of Christianity, the monastic stress on the value of manual labor, the West's emphasis on purposeful activity and a focus on individual worth which saw monotonous labor as unbecoming for man.

1198. White, Jr., Lynn. "The Legacy of the Middle Ages in the American Wild West." Speculum, 40 (1965), 191-202.

Argues that the frontiersmen brought with them a cultural and technological legacy reflecting a medieval European artisan and peasant heritage.

1199. White, Jr., Lynn. "Technology in the Middle Ages." Technology in Western Civilization, I. Edited by Melvin Kranzberg and Carroll W. Pursell, Jr. New York: Oxford University Press, 1967, pp. 66-79.

Surveys medieval technology, contrasting it with that of Byzantium and Islam and pointing out medieval innovations in military technology, agriculture, transport and machine design.

1200. White, Jr., Lynn. "The Iconography of Temperantia and the Virtuousness of Technology." Medieval Religion and Technology: Collected Essays. Berkeley: University of California Press, 1978, pp. 181-204.

Traces two elements in the growth of the bourgeois ethos. First, an increase in speculative emphasis on temperance in the thirteenth and fourteenth centuries which eventually culminated with the Puritans; second, the emergence of the clock as an image of disciplined life regulated by divine wisdom. The clock symbol, spiritualized, appeared in a new iconography of temperance which confirmed the goodness of the technological enterprise. Reprinted from Action and Conviction in Early Modern Europe: Essays in Memory of E.H. Harbison. Edited by Theodore K. Rabb and Jerrold E. Seigel. Princeton: Princeton University Press, 1969, pp. 197-219.

1201. White, Jr., Lynn. "The Origins of the Coach." Medieval Religion and Technology: Collected Essays. Berkeley: University of California Press, 1978, pp. 205-216.

Argues that a primitive coach was known in eleventh-century England, although a more advanced form could be found among western Slavs by 965. Reprinted from Proceedings of the American Philosophical Society, 114 (1970), 423-431.

1202. White, Jr., Lynn. "Cultural Climates and Technological Advance in the Middle Ages." Medieval Religion and Technology: Collected Essays. Berkeley: University of California Press, 1978, pp. 217-253.

Argues that the explanation for the rapid technical development as well as for the technological attitude of western Europe lies in those assumptions of Latin Christianity which created a cultural setting in which an aggressive manipulation of nature could flourish. Reprinted from Viator, 2 (1971), 171-201.

1203. White, Jr., Lynn. "Medical Astrologers and Late Medieval Technology." Medieval Religion and Technology: Collected Essays. Berkeley: University of California Press, 1978, pp. 297-315.

Argues that the rebirth of medical astrology in the twelfth century required precise astronomical measurements which led to an improvement in the accuracy and design of astronomical instruments. Physicians were men of high social standing who often accompanied royal patrons during warfare and so developed interests in military engineering and other applications of technology and machine design. Reprinted from Viator, 6 (1975), 295-308.

1204. White, Jr., Lynn. "Medieval Engineering and the Sociology of Knowledge." Medieval Religion and Technology: Collected Essays. Berkeley: University of California Press, 1978, pp. 317-338.

Maintains that the contempt for the manual arts characteristic of antiquity was not entirely true of the M. A. Yet, although authors such as Hugh of St. Victor tried to give the mechanic arts intellectual respectability, technology flourished only outside of learned circles. Medical astrologers provide an exception, however. Reprinted from the Pacific Historical Review, 44 (1975), 1-21.

## EDUCATION: SCHOOLS AND UNIVERSITIES

### GENERAL

1205. Abelson, Paul. The Seven Liberal Arts: A Study in Mediaeval Culture. New York, 1906. Reprint. New York: AMS Press, 1972. Pp. viii + 150.

Traces the origin of the arts curriculum from antiquity. Discusses the texts used and methods employed in instruction on the trivium subjects during the pre-university and university periods. For the quadrivial arts, treats texts, methods and the effect of Arabic materials on university teaching. Concludes that the arts tradition prevailed during the medieval period although modified.

1206. Günther, Siegmund. Geschichte des mathematischen Unterrichts im deutschen Mittelalter bis zum Jahre 1525. Berlin, 1887. Reprint. Wiesbaden: Sändig, 1969. Pp. 9 + vi + 408.

Traces mathematical instruction--texts used, methods, teachers--in the quadrivial arts as well as in optics and mechanics from the time of Bede and the Carolingian age.

1207. Haskins, Charles H. The Rise of Universities. New York, 1923. Paperback reprint. Ithaca: Cornell University Press, 1979. Pp. xi + 107.

Discusses the rise of the earliest universities, the medieval professor and his teaching methods, examinations, studies and textbooks, and student life.

1208. Leff, Gordon. Paris and Oxford Universities in the Thirteenth and Fourteenth Centuries: An Institutional and Intellectual History. New York: John Wiley & Sons, 1968. Pp. 331.

Treats these universities within the context of the intellectual and doctrinal developments of the late Middle Ages focusing on the factors peculiar to each within its own urban and institutional framework.

1209. Rashdall, Hastings. The Universities of Europe in the Middle Ages. 3 vols. Edited by Frederick M. Powicke and A.B. Emden. London: Clarendon Press, 1936.

Remains the most extensive work on all aspects of medieval universities. Vol. I treats the background and origins of the university, focusing on Paris. Vol. II covers the universities of Italy, Spain, Portugal, other universities of France as well as those of Germany, Bohemia, the low countries, Scandinavia, eastern Europe and Scotland. Vol. III is devoted to Oxford and Cambridge with material on university populations and student life. The editors have enlarged and emended Rashdall's notes and have added new bibliographical materials. The notes preceding discussion of individual universities contain bibliographical information as to the histories of the universities and their documents.

1210. Riché, Pierre. Education and Culture in the Barbarian West, Sixth through Eighth Centuries. Translated by John J. Contreni. Columbia, S. C.: University of S. C. Press, 1976. Pp. xxxvii + 557.

First published as Éducation et culture dans L'Occident barbare, 6e-8e siècles in 1962. Includes material added to the third French edition of 1972. Surveys the slow and uneven decline of Roman educational traditions in various areas of barbarian Europe and the gradual replacement of the ancient system by the Christian school, a new type of educational institution which achieved its first success with the Irish and the Anglo-Saxons. Describes educational methods, both clerical and lay, and the types of materials used. Sees Charlemagne's eighth-century Renaissance as the culmination of past developments rather than a new beginning.

1211. Riché, Pierre. Les écoles et l'enseignement dans l'occident chrétien de la fin du Ve siècle au milieu du XIe siècle. Paris: Aubier Montaigne, 1979. Pp. 462.

Traces schools and educational methods in the Christian West from the early Middle Ages to the cultural awakening of Europe during the eleventh century stressing the domination of education during this period by monks and clerics. Describes types of ecclesiastical schools, their masters, students, methods of teaching and programs of study. Also discusses lay education, both of the aristocracy and commoners.

1212. Thorndike, Lynn. University Records and Life in the Middle Ages. New York, 1944. Paperback reprint. New York: W.W. Norton, 1975. Pp. xv + 476.

Presents brief excerpts on a broad variety of topics dealing with all aspects of university life from a selection of university documents and contemporary writers. Nearly half concern Paris, taken from the Chartularium universitatis Parisiensis. The passages are arranged chronologically extending into the seventeenth century. See item 1224.

## SCHOOLS AND UNIVERSITIES: SPECIFIC TOPICS

1213. Beaujouan, Guy. "L'enseignement de arithmétique élémentaire à l'université de Paris aux XIIIe et XIVe siècles." Homenaje à Millás Vallicrosa, I. Barcelona: Consejo superior de investigaciones scientíficas, 1954, pp. 93-124.

Traces the revolutionary change in the teaching of arithmetic during the thirteenth and fourteenth centuries when older techniques were replaced by new methods and new manuals of instruction took the place of older works. Illustrates the difference between the old and new, and discusses the early attitudes toward mathematical teaching at Paris when Nicomachus and Boethius were replaced by Alexander of Villedieu and Sacrobosco.

1214. Beaujouan, Guy. "Motives and Opportunities for Science in Mediaeval Universities." Scientific Change. Edited by A.C. Crombie. New York: Basic Books, 1963, pp. 219-236.

Argues that the place of science in the university went beyond the quadrivium tradition thanks to the introduction of new materials. Not all universities emphasized science; Oxford stressed mathematics, logic and optics. At Paris science teaching was not compulsory, although there as elsewhere there was private teaching of the sciences. The ideas on science formed within the university context could be creative, but remained within an Aristotelian context.

1215. Beaujouan, Guy. "La bibliothèque et l'école médicale du monastère de Guadalupe à l'aube de la Renaissance." Médecine humaine et vétérinaire à la fin du moyen âge. Edited by Guy Beaujouan. Paris: Minard, 1966, pp. 367-468.

Discusses the hospital and the medical books in the library at Guadalupe as well as the personalities connected with the institution, primarily in the second half of the fifteenth century. Sees Guadalupe as a hospital which was transformed into a center for study and teaching based on the practical care of patients and remote from the theoretical concerns of the universities.

1216. Bullough, Vern L. "The Development of the Medical University at Montpellier to the End of the Fourteenth Century." Bulletin of the History of Medicine, 30 (1956), 508-523.

Traces the development of medical teaching at Montpellier from the first mention of medical study, c. 1137. Discusses features of the early regulations of 1220 governing masters, length of training and degree requirements, as well as those of the 1340 statutes. Surgery was not a degree subject at Montpellier and was probably taught outside the university, although Montpellier was the first French university to study anatomy by human dissection. By the end of the fourteenth century, the organization of the university was complete; the medical curriculum was based on Arabic and Greek authorities and medicine required some previous arts training.

1217. Bullough, Vern L. "The Medieval Medical University at Paris." Bulletin of the History of Medicine, 31 (1957), 197-211.

Drawing from Paris documents, chiefly the Chartularium, discusses qualifications, organization of the faculty, official texts and requirements for the degree and the medical license. Surgery was not a university subject in the thirteenth and fourteenth centuries. By the end of the fourteenth century, Paris was an important center for medical training. As it served as a model for later institutions, the exclusion of surgery was deleterious.

1218. Bullough, Vern L. "The Development of the Medical Guilds at Paris." Medievalia et Humanistica, 12 (1958), 33-40.

Argues that the attempts of the medical faculty to regulate the other medical professions--surgeons, barbers, apothecaries and herbalists--forced the latter to organize for self-protection. The resulting compartmentalization was unfortunate; it re-enforced the division between theory and practice.

1219. Bullough, Vern L. "Mediaeval Bologna and the Development of Medical Education." Bulletin of the History of Medicine, 32 (1958), 201-215.

Describes medical teaching at Bologna from the thirteenth century through the early fifteenth. At Bologna, in contrast to Paris and Montpellier, surgery was included in the curriculum, and by the beginning of the fourteenth century, anatomy was studied by human dissection. Drawing from the 1405 statutes, discusses the curriculum, teaching methods, examinations, salaries of physicians and qualifications for the doctor's degree.

1220. Bullough, Vern L. "Training of the Nonuniversity-Educated Medical Practitioners in the Later Middle Ages." Journal of the History of Medicine and Allied Sciences, 14 (1959), 446-458.

Discusses the education received by surgeons, barber-surgeons and pharmacists in France and England. These practitioners were trained by apprenticeship and their education was more practical than that found at the university. However, the learned tradition played a role, especially in the case of surgeons.

1221. Bullough, Vern L. "The Teaching of Surgery at the University of Montpellier in the Thirteenth Century." Journal of the History of Medicine, 15 (1960), 202-204.

Discusses a commentary published by Sudhoff which reveals that surgery was taught at Montpellier, although not formally, and that the hospital at Montpellier was used for demonstrations. See item 1026, pp. 297-384.

1222. Bullough, Vern L. "Medical Study at Medieval Oxford." Speculum, 36 (1961), 600-612.

Traces the development of the medical faculty at Oxford. It was always small, never influential and was subordinate to the arts faculty. While there is little by way of statutory regulation of medicine at early Oxford, there were several physicians connected with the school and semi-medical groups around Oxford. By the beginning of the fourteenth century, there is evidence of degrees in medicine and requirements for the practitioner. The curriculum at Oxford was similar to that elsewhere; surgery appears to have been emphasized, unlike at Paris and Montpellier. At Oxford, there was not the specialization found elsewhere; many students took medical courses who were not prospective physicians, while medical degrees were received by students who did not practice medicine subsequently.

1223. Bullough, Vern L. "The Mediaeval Medical School at Cambridge." Mediaeval Studies, 24 (1962), 161-168.

Traces the development of medical studies at Cambridge. By mid-fourteenth century, there were students at Cambridge with medical interests and medical instruction was given. Extant statutes of the late fourteenth century provide evidence that the Cambridge medical curriculum lagged behind that of both the continental universities and Oxford.

1224. Denifle, H. and E. Chatelain. Chartularium universitatis Parisiensis. 4 vols. Paris, 1889-1897. Reprint. Brussels: Editions culture et civilisation, 1964.

University documents covering the period from 1200 to 1450 with extensive notes.

1225. Evans, Gillian R. "The Influence of Quadrivium Studies in the Eleventh- and Twelfth-Century Schools." Journal of Medieval History, 1 (1975), 151-164.

Shows evidence for the study of the quadrivium and treats the sources of contemporary interest in the mathematical arts, with special reference to the quadrivium texts of Boethius and Martianus Capella in relation to the thought of William of Conches, Peter Abelard and others.

1226. Evans, Gillian R. "The Development of Some Textbooks on the Useful Arts, c. 1000-c. 1250." History of Education, 7 (1978), 85-94.

Points out that a concern in this period with the usefulness of learning led to the writing of textbooks designed for instruction in the practical aspects of the arts. These included grammatical colloquies which provided a basic knowledge of Latin, and treatises on the arts of letter-writing, poetry and preaching. The practical side of the mathematical arts of the quadrivium taught practical arithmetic by the abacus, and later by algorism, while practical geometry gave instruction in surveying with instruments in the field.

1227. Evans, Gillian R. "Schools and Scholars: The Study of the Abacus in English Schools c. 980-c. 1150." English Historical Review, 94 (1979), 71-89.

Argues that the abacus was enthusiastically studied in England. Arriving there in the tenth century, by the eleventh, there were several centers of learning where the abacus, despite its practical applications, was taught as a theoretical subject within the framework of the quadrivium.

1228. Gabriel, Astrik L. "Metaphysics in the Curriculum of Studies of the Mediaeval Universities." Garlandia: Studies in the History of the Mediaeval University. Frankfort: Josef Knecht, 1969, pp. 201-209.

Investigates the use of the Metaphysics in the arts curriculum, the length of the courses on it and the fees paid, as given in the statutes of several universities.

1229. Gibson, Margaret T. "The Artes in the Eleventh Century." Arts libéraux et philosophie au moyen âge. Actes du quatrième congrès international de philosophie médiévale. Montreal: Institut d'études médiévales, 1969, pp. 121-126.

Traces the fortunes of the arts in the eleventh century by the location of schools in which arts were studied, the choice of texts and the uses to which a knowledge of the arts might be put.

1230. Glorieux, Palémon. La faculté des arts et ses maîtres au XIIIe siècle. Paris: J. Vrin, 1971. Pp. 552.

Presents a catalogue in three parts concerning arts masters at thirteenth-century Paris:

1. A list of masters and their works on arts disciplines.
2. A chronological list of masters with no known works on arts subjects.
3. Anonymous arts treatises which cannot be attributed to known masters.

1231. Grabmann, Martin. "Eine für Examinazwecke abgefasste Quaestionensammlung der Pariser Aristenfakultät aus der ersten Hälfte des XIII. Jahrhunderts." Mittelalterliches Geistesleben, II. Munich, 1936. Reprint. Hildesheim: Olms, 1975, pp. 183-199.

Discusses the content of a collection of questions with the texts to be read compiled by an author who wished to order the disciplines covered in the examinations for the baccalaureate and the license. The classification follows the Aristotelian-Boethian division of the sciences. The collection throws light on the curriculum and texts required at Paris; the greater part of the collection is devoted to philosophia rationalis (grammar, rhetoric, logic).

1232. Kibre, Pearl. The Nations in the Mediaeval Universities. Cambridge, Mass.: Mediaeval Academy of America, 1948. Pp. xi + 240.

Traces the history of these semi-autonomous associations of students and masters, their membership, organization and functions from their first development at Bologna and Paris through their adoption and transformation at universities of Italy, France, Spain and northern Europe.

1233. Kibre, Paul. "The Faculty of Medicine at Paris, Charlatanism, and Unlicensed Medical Practices in the Later Middle Ages." Bulletin of the History of Medicine, 27 (1953).

Argues that the Parisian medical faculty made a continual effort to stamp out charlatanism and unlicensed medical practice and to control apothecaries, herbalists and surgeons. However, the combined forces of the faculty and secular and religious authorities were not successful in this goal.

1234. Kibre, Paul. "Scholarly Privileges: Their Roman Origins and Medieval Expression." American Historical Review, 59 (1954), 543-567.

Traces the development of privileges and immunities for scholars and teachers from their origin in Roman law, through the affirmation of imperial protection within the German empire by Frederick Barbarossa, to the further extension of various forms of special protection under the auspices of civil and papal authority in the university centers of France and England in the late Middle Ages.

1235. Kibre, Pearl. "Academic Oaths at the University of Paris in the Middle Ages." Essays on Medieval Life and Thought in Honor of Austin Patterson Evans. Edited by John H. Mundy, Richard W. Emery and Benjamin N. Nelson. New York: Columbia University Press, 1955, pp. 123-137.

Argues that the taking of oaths to uphold the statutes and privileges of the university was the chief sanction of the corporation of masters and students; violations were severely punished. While the university required oaths from important outsiders, such promises by university members to external agencies was discouraged. By the fifteenth century, university independence was largely over; the community could not count on the continued loyalty of its former members.

1236. Kibre, Pearl. Scholarly Privileges in the Middle Ages. Cambridge, Mass.: Mediaeval Academy of America, 1962. Pp. xvi + 446.

Studies the rights, privileges and immunities granted to medieval scholars with reference to Bologna, Padua, Paris and Oxford. Establishes that these privileges, originating in Roman and canon law and in the Authentica habita of Frederick Barbarossa (1158), were still preserved although modified at the end of the fifteenth century.

1237. Kibre, Pearl. "The Quadrivium in the Thirteenth-Century Universities, with Special Reference to Paris." Arts libéraux et philosophie au moyen âge. Actes du quatrième congrès international de philosophie médiévale. Montreal: Institut d'études médiévales, 1969, pp. 175-191.

Attests through evidence taken from thirteenth-century scholars and masters associated with the university that the quadrivial arts constituted an integral part of the Parisian arts curriculum; discusses the texts employed.

1238. Kibre, Pearl. "Arts and Medicine in the Universities of the Later Middle Ages." The Universities in the Late Middle Ages. Edited by Jozef IJsewijn and Jacques Paquet. Louvain: Louvain University Press, 1978, pp. 213-227.

Argues that the link between arts and medicine and the notion of the physician as a man of learning as well as a practitioner, stemming from antiquity, was maintained throughout the medieval period. This is confirmed by an examination of the statutes of the medical faculty at Paris and of those of the universities of Cologne and Louvain, which were modeled on Paris.

1239. Kibre, Pearl. "The Boethian De institutione arithmetica and the Quadrivium in the Thirteenth-Century University Milieu at Paris." Boethius and the Liberal Arts. Edited by Michael Masi. Berne: Lang, 1981, pp. 67-80.

Points out the long continuity of interest in and influence of Boethius' Latin paraphrase of the De institutione arithmetica of Nicomachus. The work was used at monastic and cathedral schools, and as a text in the teaching of the quadrivial arts within the Paris curriculum. It was directly used in the mathematical treatises of Sacrobosco and Jordanus de Nemore. Its philosophical orientation was attractive to thirteenth-century scholastics concerned with the importance of mathematical learning (Grosseteste and Roger Bacon), and it was drawn on by thirteenth century philosophers at Paris or connected with scholarly circles there.

1240. Kibre, Pearl. "Logic and Medicine in Fourteenth-Century Paris." Studi sul XIV secolo in memoria Anneliese Maier. Edited by Alfonso Maierù and Agostino Paravicini Bagliani. Rome: Edizioni di storia e letteratura, 1981, pp. 415-420.

Confirms the long-standing association between logic and medicine at Paris in the arts program required for future physicians. Traces the relation of logic and medicine revealed in the writings of Pietro d'Abano, Bartholomaeus of Bruges, Petrus of Sancto Floro as well as in collections of questiones.

1241. Lemay, Richard. "The Teaching of Astronomy in Medieval Universities, Principally at Paris in the Fourteenth Century." Science, Medicine and the University: 1200-1550. Essays in Honor of Pearl Kibre, II. Edited by Nancy G. Siraisi and Luke Demaitre. Manuscripta, 20 (1976), pp. 197-217.

Argues that there is considerable evidence to show that astronomy-astrology was encouraged in connection with medical studies not only at Paris, but at other universities, especially Italian. At Paris, astrology was a focus of attention by arts students and those in the medical and theological faculties, sanctioned by civil and church authority, from the thirteenth through the fifteenth centuries.

1242. Lindgren, Uta. Gerbert von Aurillac und das Quadrivium: Untersuchung zur Bildung im Zeitalter der Ottonen. Sudhoffs Archiv: Zeitschrift für Wissenschaftsgeschichte. Beiheft 18. Wiesbaden: Steiner, 1976. Pp. 125.

Outlines the career, education and the contributions of Gerbert to the study of the quadrivial arts as of the tenth and eleventh centuries. Gerbert wrote on all four subjects of the quadrivium; he was pivotal in the transmission of the abacus and a form of Arabic numerals to the West. His astronomical interests are well-attested; he constructed instruments for astronomical observation, and, through Lupitus of Barcelona, was a factor in the transmission of Arabic astronomical materials beyond the Pyrenees.

1243. Little, A.G. "The Franciscan School at Oxford in the Thirteenth Century." Archivum Franciscanum Historicum, 19 (1926), 803-874.

Traces the origin of the school, Grosseteste's lectureship and that of his successors, and its relationship with the Order, the English Province and the university. Lists the Franciscans who lectured at Oxford as masters or bachelors.

1244. Maier, Anneliese. "Internationale Beziehungen an spätmittelalterlichen Universitäten." Ausgehendes Mittelalter: Gesammelte Aufsätze zur Geistesgeschichte des 14. Jahrhunderts, II. Rome: Edizioni di storia e letteratura, 1967, pp. 317-334.

Argues that the extraordinarily rapid transmission of new ideas within university circles over great distances and despite political and material difficulties can be explained by personal exchanges by individuals who carried new notions from one studium to another. Provides an unusually clear example of such an exchange in the transmission of Averroism from Oxford (William of Alnwick) and Paris (Burley) to Bologna in the early fourteenth century. Reprinted from Beiträge zum ausländischen öffentlichen Recht und Völkerrecht, 29 (1954), 205-221.

1245. Markowski, Mieczysław. "Die neue Physik an der Krakauer Universität im XV. Jahrhundert." Antiqui und Moderni: Traditionsbewusstsein und Fortschrittsbewusstsein im späten Mittelalter. Edited by Albert Zimmermann. Berlin: de Gruyter, 1974, pp. 501-508.

Maintains that the first Cracow arts masters, coming mostly from Prague, brought with them a familiarity with the doctrines of Buridan and his school, and that these authors (especially Buridan) remained a dominant influence in Cracow arts teaching until mid-fifteenth century. However, Cracow masters, in their own commentaries, reveal ideas independent of Aristotle and their medieval authorities.

1246. Markowski, Mieczysław. "Astronomie an der Krakauer Universität im XV. Jahrhundert." The Universities in the Late Middle Ages. Edited by Jozef IJsewijn and Jacques Paquet. Louvain: Louvain University Press, 1978, pp. 256-275.

Traces the growing attention given to astronomical studies at Cracow from the early fifteenth century. The endowment of two chairs for astronomical lectures, the familiarity of Cracow masters with the work of French, Italian and Viennese authors, and the mounting reputation of Cracow astronomers assured that by the last quarter of the fifteenth century, Cracow was an international center for astronomical learning.

1247. Pazzini, Adalberto. "Ancora sull'insegnamento della medicina nell'altro medioevo." Pagine di storia della medicina, 16 (1972), 24-67.

Sketches the kind of medical instruction offered in the state and ecclesiastical schools of the late Roman empire and the early Middle Ages. Provides a bibliographical essay on works of recent scholarship on medical instruction in these schools.

1248. Schrader, Dorothy V. "The Arithmetic of the Medieval Universities." Mathematics Teacher, 60 (1967), 264-278.

Discusses the sources available for the study of arithmetic from the early M. A. through the university period.

1249. Seidler, Eduard. Die Heilkunde des ausgehenden Mittelalters in Paris: Studien zur Struktur der spätscholastischen Medizin. Sudhoffs Archiv: Vierteljahrsschrift für Geschichte der Medizin und der Naturwissenschaften, der Pharmazie und der Mathematik. Beiheft 8. Wiesbaden: Steiner, 1967. Pp. xvi + 162.

Treats the structure and organization of the medical faculty at Paris in the fourteenth century, the available medical works as revealed in MSS in Parisian libraries, the impact of contemporary philosophical issues on medical teaching and the contributions of Parisian masters.

1250. Siraisi, Nancy G. Arts and Science at Padua: The Studium of Padua before 1350. Toronto: Pontifical Institute of Mediaeval Studies, 1973. Pp. 199.

Analyzes the curriculum and structure of Padua so as to determine the relationship between the arts, especially astronomy-astrology, natural philosophy and medicine at Padua before 1350.

1251. Siraisi, Nancy G. "The libri morales in the Faculty of Arts and Medicine at Bologna: Bartolomeo da Varignana and the Pseudo-Aristotelian Economics." Manuscripta, 20 (1976), 105-118.

Points out the relevance of the libri morales (Ethics, Politics, Economics) as seen in the Economics commentary by Bartolomeo written for teaching purposes. Concludes that the Economics provided an opportunity to treat practical problems of social and economic life important to professional physicians.

1252. Thomson, Ron B. "Jordanus de Nemore and the University of Toulouse." British Journal for the History of Science, 7 (1974), 163-165.

Argues that the association between Jordanus and the University of Toulouse, first introduced by Curtze in 1887, is not valid.

1253. Weisheipl, James A. "Curriculum of the Faculty of Arts at Oxford in the Early Fourteenth Century." Mediaeval Studies, 26 (1964), 143-185.

Traces the course of studies followed by Oxford students from matriculation to regent master in arts; treats the scholastic method of teaching, the types of lectures and disputations, the texts studied and the requirements for the degrees of bachelor and master.

1254. Weisheipl, James A. "Developments in the Arts Curriculum at Oxford in the Early Fourteenth Century." Mediaeval Studies, 28 (1966), 151-175.

Maintains that logic and natural philosophy dominated the fourteenth-century Oxford arts curriculum. While logical works by and based on Aristotle were still valued, treatises on new logical themes were written and studied. Likewise, while the *libri naturales* remained basic, Oxford masters introduced a new logical and mathematical analysis of physical problems characteristic of the Merton school.

1255. Weisheipl, James A. "The Place of the Liberal Arts in the University Curriculum during the Fourteenth and Fifteenth Centuries." *Arts libéraux et philosophie au moyen âge. Actes du quatrième congrès international de philosophie médiévale*. Montreal: Institut d'études médiévales, 1969, pp. 209-213.

Maintains that the curriculum in arts in the early fourteenth century included all the arts as well as the three philosophies, although after the mid-fourteenth century the number of years necessary to complete the course as well as the number of required texts was reduced. The revival of interest in letters and humanities at the end of the Middle Ages took place apart from university circles.

1256. Weisheipl, James A. "The Parisian Faculty of Arts in the Mid-Thirteenth Century: 1240-1270." *American Benedictine Review*, 25 (1974), 200-217.

Summarizes the course of study in arts for undergraduates, bachelors and masters, the texts used and the teaching methods (lecture and disputation) at Paris during a period of rapid development and change.

1257. Wickersheimer, Ernest. *Commentaires de la faculté de médicine de l'université de Paris (1395-1516)*. Paris: Imprimerie nationale, 1915. Pp. xcvii + 555.

An edition of the first three volumes of the *Commentaires* of the medical faculty at Paris, a collection of records of faculty business. Introduces the edition with a discussion of the origin of the medical faculty, the books in its library (1395-1516), its organization and administration, as well as the struggles of the faculty with ancillary professionals and non-authorized practitioners.

## SALERNO AND SALERNITANS

1258. Anderson, Frank J. "New Light on Circa instans." Pharmacy in History, 20 (1978), 65-68.

Points out that the Circa instans was a twelfth-century product of Salerno and the first major attempt of western pharmacy to go beyond earlier work of the Greeks and Romans.

1259. Bayon, H.P. "The Masters of Salerno and the Origins of Professional Medical Practice." Science, Medicine and History: Essays on the Evolution of Scientific Thought and Medical Practice Written in Honour of Charles Singer, I. Edited by E. Ashworth Underwood. London: Oxford University Press, 1953, pp. 203-219.

Argues in support of a medical professionalism among the masters of Salerno before any evidence of an organized school.

1260. Corner, George W. Anatomical Texts of the Earlier Middle Ages. A Study in the Transmission of Culture with a Revised Latin Text of Anatomia Cophonis and Translations of Four Texts. Washington, D.C.: Carnegie Institution, 1927. Pp. 112.

Provides texts and/or translations of four anatomical works from twelfth- and thirteenth-century Salerno. Text and translation of the Anatomia porci (Cophonis), translations of the second Salernitan anatomical demonstration, of the Anatomia Magistri Nicolai physici and the Anatomia Ricardi Anglici.

1261. Corner, George W. "Salernitan Surgery in the Twelfth Century." The British Journal of Surgery, 25 (1937-38), 84-99.

Discusses the background (oral tradition, Byzantine texts in Latin, translations of Constantine the African) of the Bamberg Surgery and the Surgery of Master Roger Frugardi, surveying their content and giving illustrative passages.

1262. Dales, Richard C. "Marius On the Elements and the Twelfth-Century Science of Matter." Viator, 3 (1972), 191-218.

Analyzes the content of Marius' work in which this twelfth-century author, who may have been a master at Salerno, presents the nature of the elements, their associated qualities and their role in compound bodies. His approach is completely naturalistic and mechanical, and in his treatment of mixtures, quantitative. While he may have drawn directly or otherwise on Aristotle, his primary source was the Pantegni of Constantine the African. See items 1268, 1281.

1263. Hurd-Mead, Kate Campbell. "Trotula." Isis, 14 (1930), 349-367.

Evaluates the work of Salerno's best known woman physician. Modern scholarship both defends and denies her existence.

1264. Kristeller, Paul O. "The School of Salerno: Its Development and its Contribution to the History of Learning." Studies in Renaissance Thought and Letters. Rome: Edizioni di storia e letteratura, 1956, pp. 495-551.

Traces the school of Salerno from its origins around the end of the tenth century to beyond the medieval period. Salerno began as an association of physicians whose teaching was practical in purpose although instruction became increasingly theoretical after the mid-eleventh century. During the twelfth century, the highpoint of the school's medieval career, a regular curriculum developed using standard texts which were the focus of scholastic commentary, and a close tie was forged between natural philosophy and medicine. By the thirteenth century, the school had attained official status and royal privileges. An organized corporation of physicians appeared only in the latter part of the fifteenth century. Reprinted from Bulletin of the History of Medicine, 17 (1944-45), 346-374.

1265. Kristeller, Paul O. "Beitrag der Schule von Salerno zur Entwicklung der scholastischen Wissenschaft im 12. Jahrhundert." Artes liberales: Von der antiken Bildung zur Wissenschaft des Mittelalters. Edited by Joseph Koch. Leiden: Brill, 1959, pp. 84-90.

Argues that the school of Salerno, beginning in the tenth century as an association of practical physicians, revealed a theoretical and philosophical orientation by the twelfth century. Medical instruction was carried on by the scholastic methods of the later universities. The connection in the medieval period in the Italian universities between medicine and natural philosophy thus can be traced back to Salerno.

1266. Lawn, Brian. The Salernitan Questions: An Introduction to the History of Medieval and Renaissance Problem Literature. Oxford: Clarendon Press, 1963. Pp. xiv + 240.

Traces the origin of and the long tradition established by the collection of questions used in scientific and medical teaching at Salerno. The questions were the work of several masters, and from around 1200 spread throughout the West where the question literature continued to multiply through the seventeenth century.

1267. Lawn, Brian, ed. The Prose Salernitan Questions. London: Oxford University Press, 1979. Pp. xxviii + 416.

An edition of a collection of prose questions and answers, Salernitan in origin and spirit, anonymous, but of English origin, written about 1200, taken from Bodleian MS Auct. F. 3. 10. Appendixes contain editions of 10 related collections of questions.

1268. Marius. On the Elements. Edited with English translation by Richard C. Dales. Berkeley: University of California Press, 1976. Pp. 206.

See items 1262, 1281.

1269. Nutton, Vivian. "Velia and the School of Salerno." Medical History, 15 (1971), 1-11.

Argues against the thesis that there was a medical school at Velia which was transferred to Salerno. There were doctors and philosophers at early Velia, but no medical school. A tradition of classical medicine was still alive in Italian cities in the early medieval period; several may have influenced Salerno.

1270. O'Neill, Ynez Violé. "Another Look at the Anatomia porci." Viator, 1 (1970), 115-124.

Reviews the significant literature concerning this work, examines theories concerning it, and suggests a few possible sources.

1271. Regimen sanitatis Salernitanum. Edited by G. Barbensi. Florence: Olschki, 1947. Pp. x + 27.

1272. Regimen sanitatis Salernitanum. The Regimen of Health of the Medical School of Salerno. Translated with commentary by Pascal P. Parente. New York: Vantage, 1967. Pp. 93.

1273. de Renzi, Salvatore, ed. Collectio Salernitana. 5 vols. Naples, 1852-1859. Reprint. Bologna: Forni, 1967.

A collection of documents from Salerno.

1274. Rowland, Beryl. "Exhuming Trotula, sapiens matrona of Salerno." Florilegium, 1 (1979), 42-57.

Discusses gynecological treatises ascribed particularly to Trotula, focusing on an English Liber Trotularis. Trotula remains a rather unclear figure; the many works on women's ailments attributed to her cannot be by one author. Speculates on the negative connotations attached to the name "Trot," signifying an old crone whose own sexual experience is in the past, but who is expert in women's diseases, problems of childbirth, abortion and contraception. There is no firm evidence that Trotula was a medieval woman physician.

1275. Saffron, Morris Harold, ed. Maurus of Salerno, Twelfth-Century Optimus Physicus with his Commentary on the Prognostics of Hippocrates. Transactions of the American Philosophical Society, n. s., 62, 1. Philadelphia: American Philosophical Society, 1972. Pp. 104.

Provides the Latin text and translation of the commentary of this Salernitan physician on Hippocrates' treatise dealing with diseases involving acute fevers, usually fatal.

1276. Schneider, Ilse. "Die Schule von Salerno als Erbin der antiken Medizin und ihre Bedeutung für das Mittelalter." Philologus, 115 (1971), 278-291.

Argues that Salerno's ancient reputation as a spa, and its location in the former Magna Graecia, made it the obvious heir of classical medical traditions. Treats the nature of Salernitan literature and practice from the early period (pre-Constantine), through the work of Constantine the African and during the high period of Salernitan development.

1277. Sigerist, Henry E. "Bedside Manners in the Middle Ages: The Treatise *De cautelis medicorum* Attributed to Arnald of Villanova." *Henry E. Siegerist on the History of Medicine*. Edited by Félix Martí-Ibañez. New York: MD Publications, 1960, pp. 131-140.

Shows the extent to which the medieval physician had to be on guard against deceptions practiced by his patients and his need for circumspect behavior in making house calls on the sick. The work is actually an abstract of a treatise *De instructione medici* by the Salernitan physician, Archimatthaeus.

1278. Sudhoff, Karl. "Zum *Regimen sanitatis Salernitanum*." *Archiv für Geschichte der Medizin*, 7 (1913-14), 360-362; 8 (1914-15), 292-293, 352-373; 9 (1915-16), 221-249; 10 (1916-17), 91-101; 12 (1920), 149-180. Reprint. Wiesbaden: Steiner, 1964-65.

Texts and notes from the *Regimen sanitatis Salernitanum*.

1279. Sudhoff, Karl. "Die vierte Salernitaner Anatomie." *Archiv für Geschichte der Medizin*, 20 (1928), 33-50. Reprint. Wiesbaden: Steiner, 1965.

Argues that this fourth anatomy from Salerno was not designed for instruction or for demonstration, but is of a contemplative or speculative nature. It stems from the second half of the twelfth century and is possibly by Urso. Latin text.

1280. Thomasen, Anne-Liese. "Salerno und das nordische Mittelalter." *Pagina di storia della medicina*, 16 (1972), 71-81.

Discusses the transmission of Salernitan influences to the Scandinavian north, focusing on the role played by the Icelander, Ravn Sveinbjörnson (d. 1213) and the Dane Henrik Harpestreng (d. 1244), both of whom traveled extensively in southern Europe and presumably were at Salerno. Evidences of Salernitan surgical practice appear in Icelandic literature by mid-twelfth century, and by 1200, Salernitan methods in the treatment of wounds and broken bones can be found in Iceland and Norway. Ravn himself followed Salernitan methods of cautery, blood-letting and treatment for stone. The Danish physician, Harpestreng, composed a herbal based on Salernitan materials which was widely used in northern lands.

1281. Thomson, Rodney M. "Liber Marii de elementis: The Work of a Hitherto Unknown Salernitan Master?" Viator, 3 (1972), 179-189.

Adduces evidence to indicate that Marius was a physician at Salerno. A copy of his On the Elements in the British Library in a MS which contains other Salernitan materials indicates that work from Salerno was in England from 1175. See items 1262, 1268.

1282. Trotula of Salerno. The Diseases of Women by Trotula of Salerno. Translated by Elizabeth Mason-Hohl. Los Angeles: Ward Ritchie Press, 1940. Pp. x + 52.

A translation of the De passionibus mulierum curandorum. Argues that Trotula, a woman physician, practiced and taught medicine at eleventh-century Salerno.

1283. Tuttle, Edward F. "The Trotula and Old Dame Trot: A Note on the Lady of Salerno." Bulletin of the History of Medicine, 50 (1976), 61-72.

Discusses the evidence for the existence of Trotula, possibly legendary Salernitan physician and author.

1284. Urso of Salerno. De commixtionibus elementorum libellus. Edited by Wolfgang Stürmer. Stuttgart: Klett, 1976. Pp. 230.

## CHARTRES AND CHARTRIANS

1285. Bernard Silvester. "Un manuale di geomanzia presentato da Bernardo Silvestre da Tours (XII secolo):

L'*Experimentarius*." Edited by Mirella Brini Savorelli. *Rivista critica di storia della filosofia*, 14 (1959), 283-324.

An edition of a treatise on astrological geomancy attributed to Bernard.

1286. Bernard Silvester. *Bernardi Silvestris De mundi universitate libri duo, sive Magacosmus et microcosmus*. Edited by C.S. Barach and Johann Wrobel. Innsbruck, 1876. Reprint. Frankfort: Minerva, 1964. Pp. xxi + 71.

1287. Bernard Silvester. *The Cosmographia of Bernardus Silvestris*. Translated by Winthrop Wetherbee. New York: Columbia University Press, 1973. Pp. viii + 180.

1288. Bernard Silvester. *Cosmographia*. Edited with introduction and notes by Peter Dronke. Leiden: Brill, 1978. Pp. viii + 196.

1289. Chenu, M.-D. "Nature and Man at the School of Chartres in the Twelfth Century." *The Evolution of Science*. Edited by Guy S. Métraux and François Crouzet. New York: New American Library, 1963, pp. 220-235.

Argues that the work of the masters of Chartres represented one facet of the transformation of twelfth-century society. Inspired by the *Timaeus* and other Neoplatonic sources, Chartrians revealed their part in this awakening in their insistence that nature is governed by scientific laws.

1290. Clerval, A. "L'enseignement des arts liberaux à Chartres et à Paris dans la première moitié du XIIe siècle d'après l'*Heptateuchon* de Thierry de Chartres." *Congrès scientifique international des catholiques*, II. Paris, 1889, pp. 277-296.

Describes the program of liberal arts found in the encyclopedia of arts composed during the first half of the twelfth century by Thierry of Chartres. The work lists 45 treatises covering all 7 arts. It provides information as to the state of knowledge of the sciences at that time and also evidence of the first translations from Arabic.

1291. Clerval, A. Les écoles de Chartres au moyen âge du Ve au XVIe siècle. Paris, 1895. Reprint. Geneva: Slatkine, 1977. Pp. 572.

Traces the fortunes of the schools at Chartres from the obscurities of the early period through the "golden age" under Fulbert and Ivo in the eleventh and first half of the twelfth centuries to the end of the Middle Ages. Discusses Chartrian bishops, chancellors, masters and students, the program of studies and the writings of prominent Chartrians. From the second half of the twelfth century on, Chartres was eclipsed by Paris, its schools modified in organization and personnel.

1292. Dronke, Peter. "William of Conches's Commentary on Martianus Capella." Études de civilisation médiévale (IXe-XIIe siècles). Mélanges offerts à Edmond-René Labande. Poitiers: C. E. S. C. M., 1974?, pp. 223-235.

Maintains that although William of Conches' commentary on Martianus is apparently not extant, a MS at Florence contains fragments of a commentary which likely contains excerpts from William's work.

1293. Gibson, Margaret T. "The Study of the Timaeus in the Eleventh and Twelfth Century." Pensamiento, 25 (1969), 183-194.

Argues that while the Timaeus was in ninth- and tenth-century libraries, no full-fledged commentary stems from this period. The work reached the apogee of its influence in the twelfth century with many extant MSS, commentaries and extensive marginalia. Interest in the Timaeus declined after the twelfth century.

1294. Gilson, Étienne. "La cosmogonie de Bernardus Silvestris." Archives d'histoire doctrinale et littéraire du moyen âge, 3 (1928), 5-24.

Analyzes the De mundi universitate and argues that this work, an interpretation of Genesis based on the Timaeus and Neoplatonic sources, is thoroughly Christian in outlook.

1295. Jeauneau, Edouard. *Lectio philosophorum: Recherches sur l'école de Chartres*. Amsterdam: Hakkert, 1973. Pp. xvi + 395.

A collection of previously published essays (1953-1969) covering the high point of Chartrian development during the first half of the twelfth century. The first part contains essays on Chartrian masters--Bernard and Thierry of Chartres, William of Conches and John of Salisbury. The articles in part two treat Chartrian work, especially glosses and commentaries on the *Timaeus* and Boethius.

1296. Lemay, Helen R. "Science and Theology at Chartres: The Case of the Supracelestial Waters." *British Journal for the History of Science*, 10 (1977), 226-236.

Discusses the fortunes of the Christian doctrine of the waters above the firmament among Chartrians. Thierry of Chartres provided a natural cause for the waters as did Clarenbald of Arras. William of Conches rejected them, while Bernard Silvestris returned to the authority of Scripture and the fathers.

1297. McKeon, Richard P. "Medicine and Philosophy in the Eleventh and Twelfth Centuries: The Problem of the Elements." *The Dignity of Science*. Edited by James A. Weisheipl. Washington, D.C.: Thomist Press, 1961, pp. 75-120.

Argues that before the influx of Aristotle's scientific works, philosophers such as Thierry of Chartres, Abelard, William of Conches and William of St. Thierry used the elements as ordering principles in their discussions of wider problems concerning man, the universe and science in general.

1298. MacKinney, Loren C. *Early Medieval Medicine with Special Reference to France and Chartres*. Baltimore: Johns Hopkins Press, 1937. Pp. 247.

Traces the nature of medical practice and treatment, pharmacy and surgery in France during the Merovingian and Carolingian periods and discusses medical education and practice at Chartres.

1299. McVaugh, Michael R. and Frederick Behrends. "Fulbert of Chartres' Notes on Arabic Astronomy." *Manuscripta*, 15 (1971), 172-178.

Suggests that three Parisian texts show that Fulbert had a working knowledge of the astrolabe acquired within the context of new astronomical knowledge at Chartres in the first quarter of the eleventh century, and that Ralph of Liège and Ragimbold of Cologne, traditionally associated with early specific mention of the instrument and who were Fulbert's students, may have been exposed to it at Chartres. See item 885.

1300. O'Neill, Ynez Violé. "William of Conches and the Cerebral Membranes." *Clio Medica* 2 (1967), 13-21.

Investigates the sources of the terms *dura mater*, *pia mater* and *meninges* used by William in his description of the brain in the *Philosophia mundi*.

1301. O'Neill, Ynez Violé. "William of Conches' Description of the Brain." *Clio Medica*, 3 (1968), 203-223.

Investigates the sources William used in two of his encyclopedic works: the *Philosophia mundi* and the later *Dragmaticon*. Concludes that the latter work reveals the maturation of William's concepts as well as a greater diversity in the sources he employed.

1302. Raby, F.J.E. "*Nuda natura* and Twelfth-Century Cosmology." *Speculum*, 43 (1958), 72-77.

Maintains that Macrobius' idea that nature veils her secrets in allegory was much valued at Chartres.

1303. Schipperges, Heinrich. "Die Schulen von Chartres unter dem Einfluss des Arabismus." *Sudhoffs Archiv für Geschichte der Medizin und der Naturwissenschaften*, 40 (1956), 193-210.

Assesses the role of Chartres as a conduit through which Arabic scientific materials were received in the West from the period of earliest contact as well as at the time of translations and the use of the new science by Chartrian masters.

1304. Silverstein, Theodore. "The Fabulous Cosmogony of Bernardus Silvestris." Modern Philology, 46 (1948-49), 92-116.

Provides an analysis of Bernard's De mundi universitate, especially the activities of the three chief principles--Hyle, Natura and Noys. Bernard's intent, like other Chartrians, was to adapt natural philosophy to a Christian context.

1305. Silverstein, Theodore. "Elementatum: Its Appearance among the 12th Century Cosmogonists." Mediaeval Studies, 16 (1954), 156-162.

Reports the first observed use of this word by William of Conches around 1130, who employed it in his definition of an element as the minimal and simple part of a body. The word spread to other Chartrians, and was also used by Toledan writers such as John of Spain and Dominicus Gundisalvus. It was part of the vocabulary of the high Middle Ages and known to the Renaissance.

1306. Stiefel, Tina. "Science, Reason and Faith in the Twelfth Century: The Cosmologists' Attack on Tradition." Journal of European Studies, 6 (1976), 1-16.

Argues that an overlooked aspect of the twelfth-century Renaissance is the marked interest in natural philosophy and the critical attitude toward traditional authority evidenced in the work of authors such as Adelard of Bath, William of Conches and Thierry of Chartres. Fed by a new enthusiasm for logic, stimulated by the Timaeus as well as the influx of new materials entering the Latin West, these cosmologists believed a rational understanding of the natural world was paramount, and, in the case of William and Thierry, could be extended to the account of creation in Genesis.

1307. Steifel, Tina. "The Heresy of Science: A Twelfth-Century Conceptual Revolution." Isis, 68 (1977), 347-362.

Argues based on excerpts from early twelfth-century authors such as Adelard of Bath, Thierry of Chartres and William of Conches that a revolutionary change in views of nature occurred at this time in which the natural world was conceived as susceptible to systematic and rational investigation. This development has been overlooked by modern historians and had little effect on contemporaries among whom it was ignored or treated as heresy.

1308. Stock, Brian. Myth and Science in the Twelfth Century: A Study of Bernard Silvester. Princeton: Princeton University Press, 1972. Pp. xvi + 331.

Argues that Bernard's Cosmographia, written between 1143-1148, is a complex literary work combining traditional cosmology with the awakening awareness of nature characteristic of mid-twelfth century learning and society. Divided into two parts (macrocosm and microcosm), the Cosmographia, with its allegorical figures, is a dramatic mythology whose model of the cosmos incorporates the new medical and astronomical-astrological information. Compares the work with that of other twelfth-century cosmogonists--Thierry of Chartres, William of Conches and Daniel of Morley--and concludes that not only are there considerable differences between them, but that Bernard's treatise is unique.

1309. Thierry of Chartres. Commentaries on Boethius by Thierry of Chartres and his School. Edited by Nikolaus M. Haring. Toronto: Pontifical Institute of Mediaeval Studies, 1971. Pp. 619.

Contains, among other texts, an edition of Thierry's Tractatus de sex dierum operibus.

1310. Wickersheimer, Ernest. "Textes médicaux Chartrains des IXe, Xe et XIe siècles." Science, Medicine and History: Essays on the Evolution of Scientific Thought and Medical Practice written in Honour of Charles Singer, I. Edited by E. Ashworth Underwood. London: Oxford University Press, 1953, pp. 164-176.

Lists several medical MSS, with excerpts, chiefly from Chartres.

1311. William of Conches. Glosae super Platonem. Edited by Edouard Jeauneau. Paris: J. Vrin, 1965. Pp. 358.

Latin text of William's commentary on the *Timaeus*.

1312. William of Conches. *Dragmaticon philosophiae-Dialogus de substantiis physicis*. Strassburg, 1567. Reprint. Frankfort: Minerva, 1967. Pp. 385.

## QUASISCIENCE

### ASTROLOGY

1313. Abraham ibn Ezra. The Beginning of Wisdom: An Astrological Treatise by Abraham ibn Ezra. Edited by Raphael Levy and Francisco Cantera. Johns Hopkins Studies in Romance Literatures and Languages, extra vol., 14. Baltimore: Johns Hopkins Press, 1939. Pp. 235 + lxxvi.

An edition of the Old French version of the work written originally in Hebrew and translated into French in 1273 by Hagin. Also provides the Hebrew text with English translation.

1314. Albertus Magnus. Speculum astronomiae. Edited by Stefano Caroti, Michela Pereira, and Stefano Zamponi under the direction of Paola Zambelli. Pisa: Domus Galilaeana, 1977. Pp. xiii + 210.

1315. d'Alverny, Marie-Thérèse. "Astrologues et theologiens au XIIe siècle." Mélanges offerts a M.-D. Chenu. Paris: J. Vrin, 1967, pp. 31-50.

Argues that twelfth century naturalists defended their interest in astrology by distinguishing themselves from the mathematici of antiquity. Aware of the objections of the Church Fathers to astrological fatalism, they insisted that the planets and the stars executed the divine will. The problem is represented here by the arguments of the Cistercian Helimand of Froidment and Eudes of Champagne.

1316. d'Alverny, Marie-Thérèse and Françoise Hudry. "al-Kindī, *De radiis*." *Archives d'histoire doctrinale et littéraire du moyen âge*, 41 (1974), 139-269.

An edition of the *De radiis* containing the very influential notion that the universe is permeated by a universal sympathy which relates the stars to terrestrial bodies and makes possible magical operations.

1317. Birkenmajer, Aleksander. "Pierre de Limoges, commentateur de Richard de Fournival." *Études d'histoire des sciences et la philosophie du moyen âge*. Edited by Aleksandra Maria Birkenmajer and Jerzy B. Korolec. Studia Copernicana, I. Warsaw: Polish Academy of Sciences, 1970, pp. 225-235.

Traces the discovery of a work by Peter of Limoges on the nativity of Fournival.

1318. Coopland, G.W. *Nicole Oresme and the Astrologers: A Study of his Livre de divinacions*. Cambridge, Mass.: Harvard University Press, 1952. Pp. 221.

Editions with translations of the *Livre de divinacions* and the earlier *Tractatus contra judiciarios astronomos*. Points out that while Oresme was prepared to accept certain general influences from the celestial region, he was opposed to judicial astrology based on his firm conviction that astrology could not be an exact discipline and lacked its claimed predictive power.

1319. Dales, Richard C. "Robert Grosseteste's Views on Astrology." *Mediaeval Studies*, 29 (1967), 357-363.

Argues that in his earlier work, Grosseteste accepted the basic theses of astrology. However, through the 1220's and after, he became increasingly disenchanted with astrology. About 1235, in his *Hexameron*, his final position was complete rejection, an attitude primarily due to astrology's incompatibility with Christian teaching.

1320. Kennedy, E.S. "A Horoscope of Messehalla in the Chaucer *Equatorium* Manuscript. *Speculum*, 34 (1959), 629-230.

Describes this horoscope in a manuscript which is likely a Chaucer holograph. See item 573.

1321. Kibre, Pearl. "Astronomia or Astrologia Ypocratis." Science and History: Studies in Honor of Edward Rosen. Studia Copernicana, XVI. Edited by Erna Hilfstein, Pawel Czartoryski and Frank D. Grande. Warsaw: Ossolineum, 1978, pp. 133-156.

Discusses the content and influence of this tract on astrological medicine associated with the prestigious Greek physician Hippocrates which gives forecasts of the course of health and sickness according to the position of the moon in the twelve houses or the twelve zodiacal signs. The work, probably of Greek origin, was translated three times in the late Middle Ages; it retained its authority through the sixteenth century. See item 161.

1322. Lemay, Richard. Abu Ma'shar and Latin Aristotelianism from the Twelfth Century: The Recovery of Aristotle's Natural Philosophy through Arabic Astrology. Beirut: American University, 1962. Pp. xl + 468.

Argues that Aristotelian natural philosophy and cosmology became known to the Latin West prior to the translations of the libri naturales through translations of important Arabic works on astrology such as the Introductorium of Abu Ma'shar (Albumasar). Although Albumasar never cites an Aristotelian text by name, both his cosmology and his justification of astrology as a science are based on Aristotelian material.

1323. North, J.D. "Astrology and the Fortunes of Churches." Centaurus, 24 (1980), 181-211.

Discusses the assumed connection between cyclical astronomical phenomena such as conjunctions of the superior planets, especially Saturn and Jupiter, and historical events of note as the rise of religions, important political upheavals and catastrophic natural occurrences, introduced into the Latin West by the influential Arabic astrologer, Albumasar. A widely-held belief in prognostication from conjunctions did not go unchallenged; it was attacked by Henry of Hesse, Nicole Oresme, Pierre d'Ailly and particularly by Pico della Mirandola.

1324. Oresme, Nicole. "Quaestio contra divinatores." Edited by Stefano Caroti. Archives d'histoire doctrinale et littéraire du moyen âge, 43 (1976), 201-310.

A polemic against judicial astrology disputed at Paris in 1370.

1325. Pack, Roger A. "Auctoris incerti De physiognomonia libellus." Archives d'histoire doctrinale et littéraire du moyen âge, 41 (1974), 113-138.

An edition of a treatise ascribed to Aristotle, Avicenna and Averroes on the effect of the planets on human physical and psychological characteristics, and on physiognomy.

1326. Pereira, Michela. "Sulle opere scientifiche di Raimondo Lullo. I: La nuova astronomia." Physis, 15 (1973), 40-48.

Points out that the novelty of Lull's treatise as well as that of his medical writings lies not in the subject matter itself, but in Lull's application of his combinatorial art to various areas of inquiry. In this work, the use of his art gave astrology the semblance of a rational structure.

1327. Pereira, Michela. "Ricerche interno al Tractatus novus de astronomia di Raimundo Lullo." Medioevo, 2 (1976), 169-226.

Discusses several aspects of Lull's treatise--circumstances of its composition (Paris, 1297), connections with other Lullian works, diffusion and influence of the Tractatus and its rationale. Convinced of the correspondence between the celestial and sublunar regions, Lull applied his art to the stars, planets and elements to demonstrate this astrological link.

1328. Poulle, Emmanuel. "Horoscopes princiers des XIVe et XVe siècles." Bulletin de la société nationale des antiquaires de France, (1969), 63-77.

Discusses three collections of horoscopes of royal personages and of important astronomical events, one at St. John's College, Oxford and two at the

Bibliothèque Nationale at Paris. Points out that such horoscopes can provide evidence of royal birthdates and other biographical information not available elsewhere or can correct information taken from other sources. In the case of the collections in BN, lat. MS 7443, apart from the information supplied for individuals, it is apparent that the horoscopes also provided astrological explanations for political events.

1329. Pruckner, Hubert. Studien zu astrologischen Schriften des Heinrich von Langenstein. Leipzig: Teubner, 1933. Pp. 286.

Illustrates Henry of Langenstein's aversion to astrology through editions of his Questio de cometa (1368/69) which refutes the opinions of astrologers as to the significance of the comet of 1368, and his Tractatus contra conjunctionistas which is directed against the fundamentals of astrological speculation. Also includes the Tractatus contra astrologos of Nicole Oresme who shared Henry's anti-astrological attitude.

1330. Thorndike, Lynn. "Albumasar in Sadan." Isis, 45 (1954), 22-32.

Summarizes the content of this treatise, translated into Latin, a miscellaneous collection of Arabic astrology, with anecdotes about and interrogations of Albumasar gathered by his followers. Comets are placed in the celestial region beyond Saturn.

1331. Thorndike, Lynn. "The True Place of Astrology in the History of Science." Isis, 46 (1955), 273-278.

Argues that for centuries before Newton astrology as the belief that earthly phenomena were governed by influences from the celestial region, was a generally accepted natural law.

1332. Wedel, Theodore O. The Mediaeval Attitude toward Astrology, Particularly in England. New Haven, 1920. Reprint. New York: Archon Books, 1968. Pp. 163.

Argues that from the time of Isidore to John of Salisbury, although superstitious astrology as opposed to natural was condemned, astrology was a matter of purely academic discussion. The introduction of Aristotle's works and the astrological treatises of Ptolemy and Albumasar stimulated a world view in which astrology triumphed despite condemnations of judicial astrology.

1333. Yates, Frances A. "The Art of Ramon Lull: An Approach to it through Lull's Theory of the Elements." Journal of the Warburg and Courtauld Institutes, 17 (1954), 115-173.

Treats the use of the Lullian art as applied to the heavens in Lull's Tractatus novus de astronomia. Lull assigned letters of the alphabet to the four elements, their associated qualities, and to the signs and planets depending on the latter's affinity with the elements or qualities. This elaborate scheme was employed to show various astrological influences.

1334. Zambelli, Paola. "De Aristotle a Abū Ma'shar, da Richard de Fournival a Guglielmo da Pastrengo: Un' opera astrologica controversa di Alberto Magno." Physis, 15 (1973), 375-400.

Reviews past scholarship as to the author of the Speculum astronomiae, attributed to Albertus Magnus and also to Roger Bacon. The work's mention of the Biblionomia of Fournival places it before 1260. It uses astrological material from Albumasar and was used in turn by fourteenth-century scholastics such as William of Pastrengo.

## ALCHEMY: GENERAL

1335. Berthelot, M. Histoire des sciences: La chimie au moyen âge. 3 vols. Paris: Imprimerie nationale, 1893. Pp. viii + 453 (Vol. I).

Vol. I pertains to the Latin Middle Ages. Part 1 deals with medieval traditions in chemical and alchemical technology; Part 2 treats the Latin versions of Arabic alchemical writings.

1336. Berthelot, M. Introduction à étude de la chimie des anciens et du moyen âge. Paris, 1889. Reprint. Paris: Libraire des sciences et des arts, 1938. Pp. xii + 330.

A collection of documents on the early history of chemical technology and alchemy including descriptions of apparatus, alchemical symbolism and notes on MSS and authors.

1337. Ganzenmüller, Wilhelm. Beiträge zur Geschichte der Technologie und der Alchemie. Wienheim/Berg.: Verlag Chemie, 1956. Pp. 389.

A collection of previously published essays arranged to point out the close connection between technology and alchemy. Part one contains articles on glass technology; part two deals with alchemy. Not all the essays pertain to the medieval period.

1338. Ganzenmüller, Wilhelm. Die Alchemie im Mittelalter. Paderborn, 1938. Reprint. Hildesheim: Olms, 1967. Pp. 240.

Surveys various aspects of alchemy: treatises, education of the alchemist, alchemical theory and practice, the connections between alchemy and religion and the relation of the alchemist to society.

1339. Halleux, Robert. Les textes alchimiques. Brepols: Turnhout, 1979. Pp. 153.

A survey of the origins of alchemy and various problems in the alchemical literature with a bibliography of medieval alchemy.

1340. Holmyard,E.J. Alchemy. Baltimore: Penguin Books, 1968. Pp. 288.

Surveys the history of alchemy from Greek origins through the early modern period, including developments in China and Islam. Chapters 6 through 11 are especially relevant (pp. 105-258).

1341. Kopp, Hermann. Die Alchemie in alterer und neuer Zeit. 2 vols. in 1. Heidelberg, 1886. Reprint. Hildesheim: Olms, 1962. Pp. 425.

Covers alchemy from its origins to the last quarter of the eighteenth century. The medieval section briefly discusses the alchemical "authorities"--Geber, Avicenna, Vincent of Beauvais, Albertus Magnus, Aquinas, Arnald of Villanova, Roger Bacon and Raymond Lull.

1342. von Lippmann, Edmund O. Entsehung und Ausbreitung der Alchemie. 3 vols. Berlin: Springer, 1919, 1931, Weinheim: Verlag Chemie, 1954.

Vol. I surveys Hellenistic and Arabic alchemical literature and sources. Vols. II and III are reference volumes listing alchemical and chemical substances, subjects and personalities. Vol. III was edited by Richard von Lippmann.

1343. Multauf, Robert P. The Origins of Chemistry. London: Oldbourne, 1966. Pp. 412.

Discusses alchemy in antiquity, Islam and the Latin West within the context of the history of chemistry. See chapters 5 through 9 (pp. 82-200).

1344. Ploss, Emil Ernst, Heinz Roonen-Runge, Heinrich Schipperges and H. Buntz. Alchemia, Ideologie und Technologie. Munich: Heinz Moss, 1970. Pp. 227.

Surveys European alchemy by national areas including mystical-hermetic, theological and technical aspects as well as the transmission and reception of Arabic alchemy in the Latin West.

1345. Read, John. Prelude to Chemistry: An Outline of Alchemy, its Literature and Relationships. London, 1936. Paperback reprint. Cambridge, Mass.: M. I. T. Press, 1966. Pp. xxiv + 328.

A survey of alchemy, its origins, literature and principal doctrines.

1346. Taylor, F. Sherwood. The Alchemists: Founders of Modern Chemistry. New York, 1949. Paperback reprint. New York: Collier Books, 1962. Pp. 192.

A short and straightforward account of alchemy, its goals and contributions to the science of matter. Includes Greek, Islamic and post-medieval developments.

Chapters 8 through 10 focus on medieval European alchemy (pp. 82-150).

## ALCHEMY: SPECIFIC TOPICS

1347. Albertus Magnus. Libellus de alchimia Ascribed to Albertus Magnus. Translated by Sister Virginia Heines with a foreword by Pearl Kibre. Berkeley: University of California Press, 1958. Pp. xxii + 79.

Characterizes the work as a collection of clear-cut recipies dealing with the properties of substances and chemical procedures based on personal observation which reveals a belief in the possibility of the transmutation of metals.

1348. Allard, Guy H. "Reãctions de trois penseurs du XIIIe siècle vis-à-vis de l'alchimie." La science de la nature: Theories et pratiques. Cahiers d'études médiévales, II. Paris: J. Vrin, 1974, pp. 97-106.

Maintains that the differences in response of Albert Magnus, Thomas Aquinas and Roger Bacon to alchemy does not lie in ideological factors but rather in logical and epistemological divergences.

1349. Anawati, Georges C. "Avicenne et l'alchemie." Oriente e occidente nel medioevo: Filosofia e scienze. Rome: Accademia nazionale dei Lincei, 1971, pp. 285-341.

Argues from an analysis of four alchemical texts attributed to Avicenna that three are not authentic, and confirms Ruska's opinion that Avicenna did not believe in the transmutation of metals. Suggests that the alchemical doctrine in the genuine De congelatione et conglutinatione lapidum can be explained by assuming it to be an early work in which Avicenna wished to test the allegations of the alchemists. See item 1379.

1350. Aquinas, Thomas. Aurora consurgens: A Document Attributed to Thomas Aquinas on the Problem of Opposites in Alchemy. Edited with commentary by Marie-Louise von Franz. Translated by R.F.C. Hull and A.S.B. Glover. New York: Pantheon Books, 1966. Pp. xv + 555.

Attributes this work on religious and mystical alchemy to Aquinas, presumably written during his final illness.

1351. Brehm, Edmund. "Roger Bacon's Place in the History of Alchemy." Ambix, 23 (1976), 53-58.

Concludes that Bacon's role in alchemy has been exaggerated by modern scholars. However, Bacon's formulation of the relationship between alchemy and the elixir, and Christian morality and salvation, is an important link between the ancient soteriological tradition of alchemy and the development of the art in Europe during the fourteenth century.

1352. Crisciani, Chiara. "The Conception of Alchemy as Expressed in the Pretiosa Margarita Novella of Petrus Bonus of Ferrara." Ambix, 20 (1973), 165-181.

Argues that alchemy was a genuine science to Bonus which, however, ultimately depended on divine revelation. His work thus is an alchemy that is both scientific (exoteric) and religious (esoteric).

1353. Crisciani, Chiara. "La 'Quaestio de alchimia' fra duecento e trecento." Medioevo, 2 (1976), 119-168.

Argues that the various reactions to alchemy found in Islam were reflected in the Latin West with the transmission of Arabic authorities, although western alchemy, having mastered the Arabic tradition, displayed a character of its own. This is revealed in several works in the question genre.

1354. Darmstaedter, Ernst. "Liber misericordiae Geber, eine lateinische Übersetzung der grosseren kitāb al-rahma." Archiv für Geschichte der Medizin, 18 (1925), 181-197.

The text of the Latin translation of a work by the Arabic alchemist, Jābir ibn Ḥayyān.

1355. Flamel, Nicolas. Le livre des figures hiéroglyphiques. Le sommaire philosophique. Le désir désire. With an introduction by René Alleau and a historical study of Nicolas Flamel by Eugène Canseliet with review of old editions, glossary and bibliographical notes by Maxime Préaud. Paris: Denoël, 1970. Pp. 230.

1356. Flamel, Nicolas. Alchemical Hieroglyphics. Translated from French in 1624 by Eirenaeus Orandus. Gillette, N. J.: Heptangle Books, 1980. Pp. xxi + 89.

1357. Gagnon, Claude. "Recherche bibliographique sur l'alchimie médiévale occidentale." La science de la nature: Théories et pratiques. Cahiers d'études médiévales, II. Paris: J. Vrin, 1974, pp. 155-199.

Presents a rationale (with outline) and preliminary illustrative entries for a bibliography of medieval alchemy in the West.

1358. Gagnon, Claude. "Alchimie, techniques et technologie." Cahiers d'études médiévales, VII. Les arts mécaniques au moyen âge. Edited by Guy H. Allard and Serge Lusignan. Paris: J. Vrin, 1982, pp. 131-146.

Analyzes two traditions in alchemy, the theoretical (or technological) and the practical so as to speculate why this branch of the mechanic arts had no subsequent development. Alchemy was ill-served by both a false physical/metaphysical base and by a lack of sufficient technical capability. Alchemy became progressively allegorized, a process which served to fill the conceptual gaps engendered by an inadequate physical theory.

1359. Ganzenmüller, Wilhelm. "Ein alchemistische Handschrift aus der zweiten Hälfte des 12. Jahrhunderts." Sudhoffs Archiv für Geschichte der Medizin und der Naturwissenschaften, 39 (1955), 43-55.

Gives the text and translation of a short alchemical work from the second half of the twelfth century, pointing out that previously published alchemical writings go back with certainty only to the beginning of the thirteenth century. The work is part of a larger one and seems to contain directions for a beginner in the art; it is dependent on Rhazes.

1360. Geber. The Works of Geber. Translated by Richard Russell. Reissued with an introduction by E.J. Holmyard. London: J. M. Dent, 1928. Pp. xxiii + 264.

A reissue of the translation made in 1678 of The Sum of Perfection and other alchemical works which have been ascribed to the Arabic alchemist Jābir ibn Ḥayyān. It is unlikely that these works are translations of Jābir.

1361. Goltz, Dietlinde, Joachim Telle and Hans J. Vermeer, eds. Der alchemistische Traktat von der Multiplikation von Pseudo-Thomas von Aquin. Sudhoffs Archiv: Zeitschrift für Wissenschaftsgeschichte. Beiheft 19. Wiesbaden: Steiner, 1977. Pp. vi + 173.

Discusses the sources and content of this treatise, author unknown, written in the fourteenth century. The work stems from a period when Arabic alchemy had been absorbed, and western authors were creating an independent tradition. Although the author draws from the Turba philosophorum, Geber, Avicenna and alchemical material attributed to Albertus Magnus and Arnald of Villanova, the work is not a mere compilation. The writer, dealing mostly with amalgams, was familiar with alchemical practice. The Latin text, with German translation as well as old German and Italian versions, is included.

1362. Heym, Gerard. "Al-Rāzi and Alchemy." Ambix, 1 (1938), 184-191.

Argues that Rhazes' (al-Rāzi) most important alchemical work, The Secret of Secrets, is a book of experiments although provided with a theoretical base. Rhazes was dependent on both the four-element and three-element theories; the purpose of alchemy was the improvement of base metals or stones by an elixir.

1363. Jābir ibn Ḥayyān. "Liber divinitatis de LXX." Archéologie et histoire des sciences. Edited by M. Berthelot. Mémoires de l'Academie des Sciences, 2 sér., 49 (1906), 310-363.

An alchemical work attributed to Jābir translated by Gerard of Cremona.

1364. Josten, C.H., ed. "The Text of John Dastin's 'Letter to Pope John XXII'." Ambix, 4 (1949), 34-51.

Text and translation of an alchemical treatise by Dastin.

1365. Kibre, Pearl. "The Alkimia minor Ascribed to Albertus Magnus." Isis, 32 (1940), 267-300.

Points out that this manual, devoid of theory, with procedures for making chemical substances, dyeing metals and producing the elixir, is part of or a supplement to the longer Semita recta or Alkimia; this latter work may be a genuine Albertian treatise. The Alkinia minor contains no mystical or allegorical elements. Includes the Latin text.

1366. Kibre, Pearl. "An Alchemical Tract Attributed to Albertus Magnus." Isis, 35 (1944), 303-316.

Maintains that despite its similarities to Albert's work on minerals, this treatise is unlikely to be genuine. It has no references to Latin authorities, but is dependent on the sulfur-mercury theory of metals associated with the author known as Geber. Transmutation is possible, but difficult.

1367. Kibre, Pearl. "The Occultis naturae Attributed to Albertus Magnus." Osiris, 11 (1954), 23-39.

Maintains that this fourteenth-century alchemical treatise, a survey based on Arabic notions, bears little resemblance to other alchemical works attributed to Albert.

1368. Kibre, Pearl. "Albertus Magnus, De occultis nature." Osiris, 13 (1956), 157-183.

Provides the text of this treatise on alchemy which reflects Arabic sources. It is very unlikely that it represents a genuine work by Albert. See item 1367.

1369. Kibre, Pearl. "Albertus Magnus on Alchemy." Albertus Magnus and the Sciences: Commemorative Essays. Edited by James A. Weisheipl. Toronto: Pontifical Institute of Mediaeval Studies, 1980, pp. 187-202.

Argues that Albert was familiar with alchemical authorities translated from Arabic and with alchemical laboratories and processes. He accepted the possibility of transmutation. Albert subsequently had a great reputation as an alchemist with many works ascribed to him. These lack the mystical elements Albert himself had deplored.

1370. Kraus, Paul. Jābir ibn Ḥayyān. 2 vols. in 1. Cairo: Imprimerie de l'institut français d'archéologie orientale, 1942, 1943.

Vol. I treats the Arabic writings by Jābir, primarily on alchemy. In Vol. II, analyzes the principal alchemical ideas in these works and their sources in Greek thought. These writings are not the work of a single individual, but rather of a school. Two were translated into Latin, the Liber de LXX and the Liber misericordiae. See items 1354, 1363.

1371. Lemay, Helen. "The Stars and Human Sexuality: Some Medieval Scientific Views." Isis, 71 (1980), 127-137.

Maintains that Arabic astrological writings, with a detailed account of how human sexuality is governed by the heavens, were highly influential in the Latin West among astrologers, medical men and natural philosophers, although these groups differed in their approach to sexual topics.

1372. Ogrinc, Will H.L. "Western Society and Alchemy, 1200-1500." Journal of Medieval History, 6 (1980), 103-137.

Surveys the fortunes of alchemy over a 300 year period within the context of the social milieu showing the changing attitudes of the various strata of society.

1373. Partington, J.R. "Albertus Magnus on Alchemy." Ambix, 1 (1937), 3-20.

Argues that Albert believed in the transmutation of metals although he was alert to the possibility of deception. He was familiar with alchemical theories from Arabic sources, but it is unlikely that he did experiments himself. Thanks to his reputation, many alchemical treatises were attributed to him.

1374. Payen, J. "Flos florum et Semita semite: Deux traités d'alchimie attribués à Arnaud de Villeneuve." Revue d'histoire des sciences, 12 (1959), 289-300.

Analyzes and compares these treatises which have been claimed as Arnald's among those circulating under his name. Confirms that the Semita semite is in large part identical with the Flos florum. The latter shows familiarity with Arabic alchemy (sulfur-mercury theory of metals). Doubts that either treatise is by Arnald. See item 918.

1375. Petrus Bonus. The New Pearl of Great Price. Edited by A.E. Waite. London, 1894. Reprint. London: V. Stuart, 1963. Pp. xi + 441.

A translation of the Pretiosa margarita novella.

1376. Petrus Bonus. Preziosa margarita novella. Edited with notes by Chiara Crisciani. Florence: La nuova Italia, 1976. Pp. 294.

See item 1375.

1377. Ruska, Julius F., ed. Tabula Smaragdina: Ein Beitrag zur Geschichte der hermetischen Literatur. Heidelberg: Winter, 1926. Pp. vii + 248.

1378. Ruska, Julius F., ed. Turba philosophorum: Ein Beitrag zur Geschichte der Alchemie. Berlin: Springer, 1931. Pp. x + 368.

1379. Ruska, Julius F. "Die Alchemie des Avicenna." Isis, 21 (1934), 14-51.

Argues that in many of his works Avicenna is revealed as hostile to alchemy; the many alchemical works attributed to him are not genuine. Discusses these pseudo-Avicennian works most of which are of Spanish-Arabic origin although some are Latin originals.

1380. Singer, Dorothea W. "The Alchemical Testament Attributed to Raymund Lull." Archeion, 9 (1928), 43-52.

Argues that the alchemical works attributed to Lull which circulated in the fifteenth century are not by the Catalan mystic. Claims that there are numerous denials in his genuine works of the possibility of transmutation. Analyzes the text of one pseudo-Lullian work, the Testamentum.

1381. Stapleton, H.E., Hidāyat Ḥusain, R.F. Azo and G.L. Lewis. "Two Alchemical Treatises Attributed to Avicenna." Ambix, 10 (1962), 41-82.

Provides the abridged Latin text of an alchemical work, the Avicennae ad Hasen regem epistola de recta, associated with Avicenna.

1382. Thomas, Phillip D. "The Alchemical Thought of Walter of Odington." Actes du XIIe congrès international d'histoire des sciences, IIIA, Paris, 1968. Paris: Blanchard, 1971, pp. 141-144.

Maintains that in his alchemical treatise, the Icocedron, Walter attempted an application of the doctrine of the intension and remission of qualities and tried to treat the nature of alchemical change and the composition of the elements quantitatively.

1383. Weyer, Jost. "Die Bedeutung des Symbols in der mittelalterlichen Alchemie." Die vielen Namen Gottes. Edited by Meinold Krauss and Johannes Lundbeck. Stuttgart: Steinkopf, 1974, pp. 277-285.

Discusses the symbolic protrayal of the spiritual aspect of alchemy as revealed by examples taken from the medieval emblem literature. Reflects the Jungian view of alchemy as a philosophical-religious-psychological process which is merely reflected in its external experimental side.

## MAGIC AND DIVINATION

1384. [pseud.] Albertus Magnus. The Book of Secrets of Albertus Magnus: Of the Virtues of Herbs, Stones and Certain Beasts: Also a Book of the Marvels of the World. Edited by Michael R. Best and Frank H. Brightman. Oxford: Clarendon Press, 1973. Pp. xlviii + 128.

A treatise on the magical virtues of substances also known as the Experimenta of Albert.

1385. d'Alverny, Marie-Thérèse. "Survivance de la magie antique." Antike und Orient im Mittelalter. Edited by Paul Wilpert with the assistance of Paul Eckert. Berlin: de Gruyter, 1962, pp. 154-178.

Points out evidence of the extraordinary persistence of ancient magic which, despite the strong negative attitude of Christianity, can be found in various medieval sources such as literary classifications of the magical arts, the formulas, invocations and cryptic abbreviations found in the margins and blank spaces in MSS and in the rituals and incantations associated with medical practices involving animals. These attest to the remarkable vitality of ancient magical notions.

1386. Bonney, Françoise. "Autour de Jean Gerson: Opinions de théologiens sur les superstitions et la sorcellerie au début du XVe siècle." Moyen âge, 77 (1971), 85-98.

Argues that the increased preoccupation during the fourteenth and fifteenth centuries with the occult, especially sorcery, is reflected in the treatises of university-based theologians such as Gerson, who condemned various superstitions, recourse to demons, witchcraft, etc. The five works discussed all condemned such practices to no avail as they continued to flourish. With the exception of Gerson, the theologians themselves were not entirely clear as to what was illicit and what not.

1387. Charmasson, Thérèse. Recherches sur une technique divinatoire: La géomancie dans l'Occident médiéval. Geneva: Droz, 1980. Pp. 400.

Describes the technique of geomancy as revealed in treatises ascribed to Hugo of Santalla, Gerard of Cremona, Bartholomew of Parma, John of Murs, Roland l'Ecrivain and others. Concludes that this form of divination, of Arabic origin, was confined to intellectual circles and to princely courts.

1388. Ferckel, Christoph. "Die Secreta mulierum und ihr Verfasser." Sudhoffs Archiv für Geschichte der Medizin und der Naturwissenschaften, 38 (1954), 267-274.

Speculates as to the authorship of this very popular treatise, often ascribed to Albertus Magnus. Concludes that the work is certainly not by Albertus. Its author appears to have been a monk, but not a student of Albertus.

1389. Hoffmann, Gerda. "Beiträge zur Lehre von der durch Zauber verursachten Krankheit und ihrer Behandlung in der Medizin des Mittelalters." Janus, 37 (1933).

Text of Constantine the African on impotence due to witchcraft. See item 1396.

1390. Pack, Roger A. "A Pseudo-Aristotelian Chiromancy." Archives d'histoire doctrinale et littéraire du moyen âge, 36 (1969), 189-241.

Provides an edition of one of the two treatises on palmistry attributed to Aristotle in MSS from the thirteenth to the fifteenth centuries. The treatise edited here provided the basis for the epitome of a Rodericus de Majoricis.

1391. Pack, Roger A. "Pseudo-Aristoteles: Chiromantia." Archives d'histoire doctrinale et littéraire du moyen âge, 39 (1972), 289-305.

An edition of one of the treatises on palmistry ascribed to Aristotle. The other was edited by Pack in 1969. See item 1390.

1392. Pack, Roger A. "Almadel Auctor Pseudonymus: De firmitate sex scientiarum." Archives d'histoire doctrinale et littéraire du moyen âge, 42 (1975), 147-181.

An edition of this unknown author of his treatise on six kinds of divination. See item 1398.

1393. Pack, Roger A. and R. Hamilton. "Rodericus de Majoricis: Tractatus ciromancie." Archives d'histoire doctrinale et littéraire du moyen âge, 38 (1971), 271-305.

An edition of a paraphrase and condensation of the pseudo-Aristotelian work on palmistry edited by Pack in 1969. See item 1390.

1394. Paschetto, Eugenia. "Linguaggio e magia nel De configurationibus di N. Oresme." Sprache und

*Erkenntnis im Mittelalter: Akten des VI. internationalen Kongresses für mittelalterliche Philosophie*, II. Berlin: de Gruyter, 1981, pp. 648-656.

Discusses Oresme's opinion as to the efficacy and causes of invocations and prayers addressed to demons within the context of Oresme's configuration doctrine which provided an explanation of how such sounds or invocations could produce particular effects. Argues that Oresme's attitude was not altogether sceptical; he believed that incantations to demons were of power either through the effect of a configuration or indirectly by the imaginative power of the invoker, and that these effects were real and dangerous.

1395. "*Picatrix:*" *Das Ziel des Weisen von Pseudo-Magrītī*. Translated into German from Arabic by Helmut Ritter and Martin Plessner. London: Warburg Institute, 1962. Pp. lxxviii + 434.

A work of Arabic origin blending philosophical doctrines, astrology, alchemy and magic known in its Latin version as *Picatrix*.

1396. Sigerist, Henry E. "Impotence as a Result of Witchcraft." *Henry E. Sigerist on the History of Medicine*. Edited by Félix Martí-Ibañez. New York: MD Publications, 1960, pp. 146-152.

Contains a translation of Constantine the African's text on impotence due to witchcraft. The connection first seems to have been due to Hincmar of Rheims. The issue was an important one in canon law dealing with permissible remarriage after annulment. See item 1389.

1397. Tannery, Paul. "Le *liber geomantie nove* d'après le manuscrit de la Laurentienné." *Mémoires scientifiques*, 4. Issued by J.-L. Heiberg. Paris: Gauthier-Villars, 1920, pp. 403-409.

Partial text of a treatise on geomancy beginning *Estimaverunt Indi*, likely translated by Gerard of Cremona not Hugo of Santalla.

1398. Thorndike, Lynn. "Alfodel and Almadel: Hitherto Unnoted Mediaeval Books of Magic in Florentine Manuscripts." Speculum, 2 (1927), 326-331; 4 (1929), 90.

Describes a work on geomancy ascribed to Alfodel of Marengi, identified as an Arabic author, Fahl b. Sahl al-Sarashi, and translated by Gerard of Cremona, whose purpose is to answer questions by the chance selection of a number. Almadel's treatise, On the Firmness of the Six Sciences, treats several forms of divination.

1399. Thorndike, Lynn. "Robertus Anglicus and the Introduction of Demons and Magic into Commentaries upon the Sphere of Sacrobosco." Speculum, 21 (1946), 241-243.

Points out that Robert introduced both astrology and divinatory spirits in his De spera commentary. Cecco d'Ascoli added occult themes such as necromancy and diabolical magic to his own commentary. See item 504, pp. 143-246.

1400. Thorndike, Lynn. "Further Consideration of the Experimenta, Speculum astronomiae and De secretis mulierum Ascribed to Albertus Magnus." Speculum, 30 (1955), 413-443.

Argues that the Experimenta and the De secretis mulierum are in part by Albert, that is, are taken from or based on his genuine writings. Likewise suggests that there is no reason not to ascribe the Speculum astronomiae to Albert.

* Thorndike, Lynn. A History of Magic and Experimental Science. 8 vols. New York: Columbia University Press, 1923-1958.

Cited as item 19. Given Thorndike's underlying assumptions concerning the interconnection between magic and science, these volumes are rich in information on the occult arts.

1401. Venancius of Moerbeke. "A Treatise on Prognostications." Edited by Roger A. Pack. Archives d'histoire doctrinale et littéraire du moyen âge, 43 (1976), 311-322.

An edition of a treatise by a Flemish author of the first half of the fifteenth century on prophetic signs observed when awake or in dreams. Part of the material was taken from a William of Aragon.

## REPORTS ON MANUSCRIPTS

1402. d'Alverny, Marie-Thérèse. "Avicenna latinus." Archives d'histoire doctrinale et littéraire du moyen âge, 28 (1961), 281-316-I; 29 (1962), 217-233-II; 30 (1963), 221-272-III; 31 (1964), 271-286-IV; 32 (1965), 257-302-V; 33 (1966), 305-327-VI; 34 (1967), 315-343-VII; 35 (1968), 301-335-VIII; 36 (1969), 243-280-IX; 37 (1970), 327-361-Supplementum; 39 (1972), 321-341-Supplementum.

A survey of Avicenna MSS.

1403. Bataillon, Louis. "Adam of Bocfeld: Further Manuscripts." Medievalia et Humanistica, 13 (1960), 35-39.

See items 1448, 1450.

1404. Birkenmajer, Aleksander. "La bibliothèque du Richard de Fournival." Études d'histoire des sciences et la philosophie du moyen âge. Edited by Aleksandra Maria Birkenmajer and Jerzy B. Korolec. Studia Copernicana, 1. Warsaw: Polish Academy of Sciences, 1970, pp. 117-210.

Traces the fortunes of Fournival's library of texts on philosophy, science, medicine, law and theology as listed in his Biblionomia. Compares the Biblionomia with the Sorbonne catalogues (many of Fournival's MSS were eventually given to the Sorbonne), and with MSS currently in the Bibliothèque Nationale which were originally in Fournival's library.

1405. Birkenmajer, Aleksander. "Eine neue Handschrift des Liber de naturis inferiorum et superiorum des Daniel von Morlai." Études d'histoire des sciences et la philosophie du moyen âge. Edited by Aleksandra Maria Birkenmajer and Jerzy B. Korolec. Studia Copernicana, 1. Warsaw: Polish Academy of Sciences, 1970, pp. 45-51.

Reports a MS at Berlin with Daniel's cosmological treatise.

1406. Boncompagni, Baldassarre. "Catalogo de'lavori di Andalò di Negro." Bullettino di bibliografia e di storia delle scienze matematiche e fisiche, 7 (1874), 339-376.

A catalogue of the extant works of Andalò, chiefly on astronomy and astrology, printed or in MSS, with editions and MSS as well as a list of treatises putatively ascribed to him cited in other works.

1407. Clagett, Marshall. "The Life and Works of Giovanni Fontana." Annali dell'Istituto e Museo di storia della scienze di Firenze, 1 (1976), 5-28.

A catalogue of the twenty works attributed to this fifteenth-century Venetian. See item 848, pt. II.

1408. Cranz, F. Edward. "The Publishing History of the Aristotle Commentaries of Thomas Aquinas." Traditio, 34 (1978), 157-192.

Traces the fortunes of the printed versions of these commentaries. Aquinas wrote expositions on a large number of Aristotelian treatises; none were printed prior to 1470. Provides a list of editions before 1650.

1409. Curtze, Maximilian. "Die Handschrift No. 14836 der Königl. Hof-und Staatsbibliothek zu München." Abhandlungen zur Geschichte der Mathematik, 7 (1895), 77-142.

Describes the content of this astronomical and mathematical MS at Munich.

1410. Dondaine, A. and L.J. Bataillon. "Le commentaire de Saint Thomas sur les Météores." Archivum fratrum praedictorum, 36 (1966), 81-152.

Discusses the confused MS tradition of the work, focusing on the continuators of the last two books. Despite a contrary Renaissance tradition, Aquinas' own commentary stopped at the penultimate chapter of Book II.

1411. Ermatinger, Charles J. "Averroism in Early Fourteenth-Century Bologna." Mediaeval Studies, 16 (1954), 35-56.

Provides a collation of MSS in published studies which have established an Averroist tradition at early fourteenth-century Bologna. Gives the location of the MSS and the problems dealt with, including materials in Vat. lat. MS 6768. See item 1436.

1412. Ermatinger, Charles J. "Notes on Some Early Fourteenth-Century Scholastic Philosophers." Manuscripta, 3 (1959), 155-168, 4 (1960), 29-38.

A bibliographical account of items in MSS and printed sources on early fourteenth-century Averroism.

1413. Ermatinger, Charles J. "A Second Copy of a Commentary on Aristotle's Physics Attributed to Siger of Brabant." Manuscripta, 5 (1961), 41-49.

Reports an additional MS at Erfurt of a Physics commentary similar to that edited by Delhaye. See item 259.

1414. Ermatinger, Charles J. "Some Unstudied Sources for the History of Philosophy in the Fourteenth Century." Manuscripta, 14 (1970), 3-33; 67-87.

Draws attention to some largely unstudied sources for philosophical thought among European arts masters. Includes John of Jandun's second treatise on the sensus agens, questions by Sebastian of Aragon, annotated by Hugo of Utrecht and Simon of Padua, other materials by Hugo and some unstudied questions on the Physics.

1415. Faral, Edmond. "Jean Buridan. Notes sur les manuscrits, les éditions et le contenu de ses ouvrages." Archives d'histoire doctrinale et littéraire du moyen âge, 15 (1946), 1-53.

Lists MSS and editions of Buridan works with question titles. Some of these treatises are now available in modern editions. See item 1416.

1416. Federici Vescovini, Graziella. "Su alcuni manoscritti di Buridano." Rivista critica di storia della filosofia, 15 (1960), 413-427.

Lists some MSS of Buridan works on logic, natural philosophy, metaphysics and ethics in Italian collections which are not given by Faral.

1417. Folkerts, Menso. "Mittelalterliche mathematische Handschriften in westlichen Sprachen in der Berliner Staatsbibliothek: Ein vorläufiges Verzeichnis." Mathematical Perspectives: Essays on Mathematics and its Historical Development. Edited by Joseph W. Dauben. New York: Academic Press, 1981, pp. 53-93.

Lists the MSS in this collection that have mathematical contents.

1418. Folkerts, Menso. "Mittelalterliche mathematische Handschriften in westlichen Sprachen in der Herzog August Bibliothek Wolfenbüttel: Ein vorläufiges Verzeichnis." Centaurus, 25 (1981), 1-49.

Lists the mathematical MSS in western languages in this important collection, a preliminary work in a project to collect all mathematical texts in various libraries.

1419. Grabmann, Martin. Studien zu Johannes Quidort von Paris, O. Pr. Sitzungsberichte der Bayerischen Akademie der Wissenschaften, philos.-philolog. und hist. Klasse, 3. Munich, 1922. Pp. 60.

Reviews the literary legacy found in the MSS of the late thirteenth-century John of Paris.

1420. Grabmann, Martin. Neu aufgefundene Werke des Siger von Brabant und Boetius von Dacien. Sitzungsberichte der Bayerischen Akademie der Wissenschaften, 2. Munich, 1924. Pp. 48.

Discusses MS versions of works by these scholastics, both connected with thirteenth-century Latin Averroism.

1421. Grabmann, Martin. Die Aristoteleskommentare des Simon von Faversham. Sitzungsberichte der Bayerischen Akademie der Wissenschaften, phil.-hist. Klasse, 3. Munich, 1933. Pp. 40.

Reports MSS versions of works by Simon of Faversham of Oxford (d. 1306) in several European collections.

1422. Grabmann, Martin. Handschriftliche Forschung und Mitteilungen zum Schriftum des Wilhelm von Conches und zu Bearbeitungen seiner naturwissenschaftlichen Werke. Sitzungsberichte der Bayerischen Akademie der Wissenschaften, phil.-hist. Klasse, 10. Munich, 1935. Pp. 57.

Contains MS information on works of William of Conches, including his Philosophia mundi, Dragmaticon philosophiae and his commentaries on Macrobius and Martianus Capella. Likewise includes such information on works that represent a reworking of William's scientific treatises.

1423. Kibre, Peal. "Hitherto Unnoted Medical Writings by Dominicus de Ragusa (1424-1425 A. D.)." Bulletin of the History of Medicine, 7 (1939), 990-995.

Reports on the works by this professor at Bologna between 1395-1427, composed while he was teaching at Siena, in a MS at the New York Academy of Medicine. See item 936.

1424. Kibre, Pearl. "Alchemical Writings Ascribed to Albertus Magnus." Speculum, 17 (1942), 499-518.

Provides a preliminary survey of MSS containing works on alchemy attributed to Albert and speculates on those written within the fourteenth century which might have a claim to authenticity. Points out that the treatises under Albert's name are free of the alchemical mysticism Albert himself deplored.

1425. Kibre, Pearl. "Further Manuscripts Containing Alchemical Tracts Attributed to Albertus Magnus." Speculum, 34 (1959), 238-247.

Adds alchemical treatises in two MSS which are ascribed to Albert and written soon after his death to the survey of 1942. See item 1424.

1426. Kibre, Pearl. "Two Alchemical Miscellanies: Vatican Latin MSS. 4091, 4092." Ambix, 8 (1960), 167-176.

Reports two MSS containing alchemical recipies as well as several works by well-known Latin and Arabic alchemical authors.

1427. Kibre, Pearl. "Hippocrates latinus: Repertorium of Hippocratic Writings in the Latin Middle Ages." Traditio, 31 (1975), 99-126, I; 32 (1976), 257-292, II; 33 (1977), 253-295, III, 34 (1978), 193-226, IV; 35 (1979), 273-302, V; 36 (1980), 347-372, VI; 37 (1981), 267-289, VII.

Lists all works known or cited in Latin under the name Hippocrates, authentic or not, with translators where known and commentators.

1428. Lattin, Harriet Pratt. "The Eleventh Century Manuscript Munich 14436: Its Contribution to the History of Coordinates, of Logic, of German Studies in France." Isis, 38 (1947-48), 205-225.

Discusses the content of this MS which contains astronomical, arithmetical and logical materials reflecting the revival in education sparked by Gerbert at the cathedral school at Rheims in the first quarter of the eleventh century. It has a tenth-century graph showing the course of the planets through the zodiac. See item 449.

1429. Lohr, Charles H. "Medieval Latin Aristotle Commentaries." Traditio, 23 (1967), 313-413, Authors A-F; 24 (1968), 149-245, Authors G-I; 26 (1970), 135-216, Jacobus-Johannes Juff; 27 (1971), 251-351, Johannes de Kanthi-Myngodus; 28 (1972), 281-396, Narcissus-Ricardus; 29 (1973), 93-197, Authors Robertus-Wilgelmus; 30 (1974), 119-124, Supplementary authors.

Lists the Latin commentaries on authentic works of Aristotle as well as commentaries on several works closely connected with Aristotle. The entries are arranged alphabetically by author's first name, with bio-bibliographical material, lists of MSS and editions. See items 1430, 1432.

1430. Lohr, Charles H. "Medieval Latin Aristotle Commentaries: Addenda et corrigenda." Bulletin de philosophie médiévale, 14 (1972), 116-126.

Presents additions and corrections to the inventory of medieval Latin Aristotle commentaries appearing in Traditio from 1967 to 1973. See item 1429.

1431. Lohr, Charles H. "Aristotelica Hispalensia." Theologie und Philosophie, 50 (1975), 547-564.

Lists MSS, primarily Aristotelian commentaries, in the Biblioteca Colombina, Sección Colombina and Sección Capitular, and in the Biblioteca Universitaria, both collections in Seville.

1432. Lohr, Charles H. "Problems of Authorship Concerning Some Medieval Aristotelian Commentaries." Bulletin de philosophie médiévale, 15 (1973, pub. 1975), 131-136.

Discusses problems concerning the authorship of several of the commentaries in the Traditio inventory. See item 1429.

1433. MacKinney, Loren C. "Medical Illustrations in Medieval Manuscripts of the Vatican Library." Manuscripta, 3 (1959), 3-18, 76-88.

1434. Maier, Anneliese. "Nouvelles Questions de Siger de Brabant sur la Physique d'Aristote." Ausgehendes Mittelalter: Gesammelte Aufsätze zur Geistesgeschichte des 14. Jahrhunderts, II. Rome: Edizioni di storia e letteratura, 1967, pp. 171-188.

Reports the discovery of a Vatican, Borghese MS containing questions on Book II of the Physics (a reportatio) attributed to Siger. These questions differ greatly from the Siger Questions on the Physics edited by Delhaye, although they reveal doctrinal similarities to Siger's other works. See item 259. Reprinted from Revue philosophique de Louvain, 44 (1946), 497-513.

1435. Maier, Anneliese. "Les commentaires sur la Physique d'Aristote attribués à Siger de Brabant." Ausgehendes Mittelalter: Gesammelte Aufsätze zur Geistesgeschichte des 14. Jahrhunderts, II. Rome: Edizione di storia e letteratura, 1967, pp. 189-206.

Points out the possibility that the commentaries on the Physics in a Vat. Borghese MS and in the edition of Delhaye are both by Siger despite their differences. Siger may have altered his views on certain doctrines. In any event, the differences between the texts is not an argument against the authenticity of the edition. See item 259. Reprinted from Revue philosophique de Louvain, 47 (1949), 334-350.

1436. Maier, Anneliese. "Die Italienischen Averroisten des Codex Vat. lat. 6768." Ausgehendes Mittelalter: Gesammelte Aufsätze zur Geistesgeschichte des 14. Jahrhunderts, II. Rome: Edizioni di storia e letteratura, 1967, pp. 350-366.

Points out that this MS contains a group of otherwise unknown questions by fourteenth-century Averroists with an index of the question titles giving the name of the author or his initials. Provides a list of the question titles and authors' names. Reprinted from Manuscripta, 8 (1964), 68-82.

1437. Markowski, Mieczysław. "Studien zu den Krakauer mittelalterlichen Physikkommentaren: Die Impetustheorie." Archives d'histoire doctrinale et littéraire du moyen âge, 35 (1968), 187-210.

Surveys commentaries on the Physics in MSS at the Jagellonian Library at the University of Cracow, with emphasis on Polish commentators on the impetus theory.

1438. Masi, Michael. "Chaucer, Messahalla and Bodleian Selden Supra 78." Manuscripta, 19 (1975), 36-47.

Suggests that Chaucer may have turned to this astronomical MS, a compilation of treatises by various authors which was in England from the mid-fourteenth century, for information used in his astronomical works and for those he intended to write.

1439. Mathieu, René. "A la recherche du De anima de Nicole Oresme." Archives d'histoire doctrinale et littéraire du moyen âge, 23 (1956), 243-252.

Discusses MS versions of Oresme's work on the De anima. A different work in Bruges MS 477 (Expositio super libros de anima) which Mathieu believes to be by Oresme may not be by him. See item 353, pp 646.

1440. Rück, Karl. "Die Naturalis historia des Plinius im Mittelalter." Sitzungsberichte der bayerischen Akademie der Wissenschaften, phil.-hist. Klasse, 1 (1898), 203-318.

Reports excerpts from Pliny in MSS at Lucca, Paris and Leiden.

1441. Schipperges, Heinrich. "Eine noch nicht veröffentlichte Summa medicinae des Petrus Hispanus in der Biblioteca Nacional zu Madrid." Sudhoffs Archiv: Vierteljahrsschrift für Geschichte der Medizin und der Naturwissenschaften, der Pharmazie und der Mathematik, 51 (1967), 187-189.

Draws attention to a MS of the thirteenth century containing ten medical works by Peter of Spain.

1442. Spade, Paul V. "Notes on some MSS of Logical and Physical Works by Richard Lavenham." Manuscripta, 19 (1975), 139-146.

Describes the MSS at Venice and in English collections containing Lavenham's work.

1443. Suter, Heinrich. "Eine bis jetzt unbekannte Schrift des Nicole Oresme." Zeitschrift für Mathematik und Physik, hist.-lit. Abteilung, 27 (1882), 121-125.

Reports the discovery of Oresme's questions on the Meteorology.

1444. Thirry, Anne. "Recherches relatives aux commentaires médiévaux du De anima d'Aristote." Bulletin de philosophie médiévale, 13 (1971), 109-128.

Presents a chronological inventory of thirteenth-century commentaries on the De anima, and other treatises on the soul and the intellect which have been edited, as well as a list of the various arguments most commonly invoked in response to the chief questions found in De anima commentaries.

1445. Thomson, Ron B. "Jordanus de Nemore: Opera." Mediaeval Studies, 38 (1976), 97-144.

A bibliography of editions of and a list of MSS of Jordanus' works including genuine treatises and dubious and spurious ascriptions.

1446. Thomson, S. Harrison. "An Unnoticed Treatise of Roger Bacon on Time and Motion." Isis, 27 (1937), 219-224.

Reports a thirteenth-century MS in Madrid containing what is apparently a Bacon work.

1447. Thomson, S. Harrison. The Writings of Robert Grosseteste, Bishop of Lincoln, 1235-1253. Cambridge: Cambridge University Press, 1940. Pp. xv + 302.

A bibliographical study of MSS materials covering Grosseteste's entire literary output, both authentic works and attributions.

1448. Thomson, S. Harrison. "A Note on the Works of Magister Adam de Bocfeld (Bochermefort)." Medievalia et Humanistica, 2 (1944), 55-87.

Confirms the identity of Bocfeld and Adam de Bouchermefort and lists MSS with works of this early commentator on Aristotle. See items 1403, 1450.

1449. Thomson, S. Harrison. "Unnoticed Questiones of Walter Burley on the Physics." Mitteilungen des Instituts für Österreichische Geschichtsforschung, 62 (1954), 390-405.

Reports the discovery of a long set of Physics questions by Burley in the University Library at Basel. Gives the titles of the questions and the text of the first question of Book III.

1450. Thomson, S. Harrison. "A Further Note on Master Adam of Bocfeld." Medievalia et Humanistica, 12 (1958), 23-32.

Lists further MSS containing Bocfeld works. See items 1403, 1448.

1451. Thorndike, Lynn. "Vatican Latin MSS in the History of Science and Medicine." Isis, 13 (1929-30), 53-102.

Treats a number of MSS; the material is organized alphabetically by author and subject.

1452. Thorndike, Lynn. "Manuscripts of the Writings of Peter of Abano." Bulletin of the History of Medicine, 15 (1944), 201-219.

Revises and enlarges the list of MSS of Peter's works beyond those given in A History of Magic and Experimental Science, II, excluding Peter's translations of Galen and astrological works.

1453. Thorndike, Lynn. "Some Little Known Astronomical and Mathematical MSS." Osiris, 8 (1948), 41-65.

Lists the contents of MSS examined at the Vatican and at Milan, Bern, Zurich, Munich and Prague.

1454. Thorndike, Lynn. "Giovanni Bianchini in Paris Manuscripts." Scripta mathematica, 16 (1950), 5-12.

Treats the work of this fifteenth-century Italian in five MSS at the Bibliothèque Nationale, including tables, canons, his Flores Almagesti and his instrument for finding the altitudes of stars.

1455. Thorndike, Lynn. "Giovanni Bianchini in Italian MSS." Scripta mathematica, 18 (1952), 5-17.

Identifies the works of this fifteenth-century astronomer in a number of Vatican MSS, and in other collections in Rome, Milan and Bologna.

1456. Thorndike, Lynn. "Oresme and Fourteenth-Century Commentaries on the Meteorologica." Isis, 45 (1954), 145-152.

Lists the questions in Oresme's commentary from Bibliothèque Nationale lat. MS 15156, and compares them with similar material in pseudo-Scotus, Themon Judaeus, Buridan and Albert of Saxony.

1457. Thorndike, Lynn. "Giovanni Bianchini's Astronomical Instrument." *Scripta mathematica*, 21 (1955), 136-137.

Gives information concerning a MS in the Bodleian containing Bianchini's instrument for finding the altitudes of celestial bodies.

1458. Thorndike, Lynn. "More Questions on the *Meteorologica*." *Isis*, 46 (1955), 357-360.

Lists the questions in the complete version of Oresme's commentary in St. Gall MS 839. See item 1456.

1459. Thorndike, Lynn. "More Dates for Late Medieval Astronomy from Some Vatican MSS." *Homenaje à Millás Vallicrosa,* II. Barcelona: Consejo superior de investigacions scientíficas, 1956, pp. 467-470.

Describes some Vatican MSS containing astronomical tables and almanachs from the late fourteenth and fifteenth centuries.

1460. Thorndike, Lynn. "Notes upon Some Medieval Latin Astronomical, Astrological and Mathematical MSS at the Vatican, Part I." *Isis*, 47 (1956), 391-404.

Describes the contents of the following MSS: Vatican Latin MSS 3098, 3099, 3126, 4082, 4085, 9410.

1461. Thorndike, Lynn. "Some Alchemical Manuscripts at Bologna and Florence." *Ambix*, 5 (1956), 85-110.

Provides extensive descriptions of MSS at the University Library, Bologna and the Laurentian, Riccardian and National Libraries at Florence.

1462. Thorndike, Lynn. "Notes on Some Astronomical, Astrological and Mathematical MSS of the Bibliothèque Nationale, Paris." *Journal of the Warburg and Courtauld Institutes*, 20 (1957), 112-172.

Describes the contents of 27 MSS, dealing with questions of authorship as well as comparisons with other MSS and printed editions.

1463. Thorndike, Lynn. "The Study of Mathematics and Astronomy in the Thirteenth and Fourteenth Centuries as illustrated by Three Manuscripts." Scripta mathematica, 23 (1957), 67-76.

Comments on the mathematical and astronomical knowledge reflected in British Library, Harley MS 3647 as compared with the more extensive contents of the later Vatican, Palatine Latin MS 1414 and Basel MS F. II. 33.

1464. Thorndike, Lynn. "Latin MSS of Works by Rasis at the Bibliothèque Nationale, Paris." Bulletin of the History of Medicine, 32 (1958), 54-67.

Gives an account of the Latin medical works by Rhazes in eight MSS at Paris; the majority are from the thirteenth century.

1465. Thorndike, Lynn. "Notes upon Some Medieval Latin Astronomical, Astrological and Mathematical MSS at the Vatican, Part II." Isis, 49 (1958), 34-49.

Describes the contents of the following MSS: Barberini 92, Ottobon 1826 and Palatine MSS 1171, 1414, 1354.

1466. Thorndike, Lynn. "Notes upon Some Medieval Astronomical, Astrological and Mathematical MSS at Florence, Milan, Bologna and Venice." Isis, 50 (1959), 33-50.

Reviews the content of five MSS from Florence, one from Bologna and two each from Milan and Venice, organized by city and collection.

1467. Thorndike, Lynn. "Some Medieval and Renaissance MSS on Physics." Proceedings of the American Philosophical Society, 104 (1960), 188-201.

Surveys the content of eight MSS from the late thirteenth century to the 1470's dealing with physical problems and located in various European collections.

1468. Weisheipl, James A. "*Repertorium Mertonense*." *Mediaeval Studies*, 31 (1969), 174-224.

A list of MSS of Merton masters between 1300-1350.

1469. Zinner, Ernst. "*Horologium viatorum*." *Isis*, 14 (1930), 385-387.

Draws attention to the many MS versions of this portable sundial extant in German collections.

1470. Zoubov, V.P. "Une fausse attribution: Le *De instantibus* attribué à Nicole Oresme." *Archives internationales d'histoire des sciences*, 11 (1958), 377-378.

Points out that the *De instantibus* in an Arsenal MS at Paris often attributed to Oresme is actually the *De instanti* of John of Holland.

INDEX